TECHNICAL ENGINEERING AND DESIGN GUIDES AS ADAPTED FROM THE US ARMY CORPS OF ENGINEERS, No. 7

BEARING CAPACITY OF SOILS

Published by
ASCE Press
American Society of Civil Engineers
1801 Alexander Bell Drive
Reston, Virginia 20191-4400

ABSTRACT

This U.S. Army Corps of Engineers engineering manual, Bearing Capacity of Soils, presents guidelines for calculation of the bearing capacity of soil under shallow and deep foundations supporting various types of structures and embankments. The principles presented are applicable to numerous types of structures such as buildings and houses, towers and storage tanks, fills, embankments, and dams. The first chapter presents definitions, failure modes, and factors that influence bearing capacity. The next chapter discusses non-load related design considerations such as frost action and soil erosion. Chapter 3 explores laboratory and in situ methods of determining soil parameters required. The last two chapters present an analysis of the bearing capacity of shallow foundations and deep foundations, respectively.

Library of Congress Cataloging-in-Publication Data

Bearing capacity of soils.
p.cm.—(Technical engineering and design guides as adapted from the US Army Corps of Engineers; no. 7)
Includes bibliographical references and index.
ISBN 0-87262-997-X
1. Foundations—United States—Testing. 2. Soils—United States—Testing. I. American Society of Civil Engineers. II. United States. Army. Corps of Engineers. III. Series: Technical engineering and design guides as adapted from the U.S. Army Corps of Engineers; no. 7.
TA775.B395 1994 94-243
624.1'5136—dc20 CIP

Library of Congress Catalog Card No: 94-243
ISBN 0-87262-997X
Manufactured in the United States of America.

TABLE OF CONTENTS

Chapter 1. Introduction

Chapter 2. Non-Load Related Design Considerations

Chapter 3. Soil Parameters

Chapter 4. Shallow Foundations

Chapter 5. Deep Foundations

DEPARTMENT OF THE ARMY
U.S. Army Corps of Engineers
WASHINGTON, D.C. 20314-1000

REPLY TO
ATTENTION OF:

Mr. James W. Poirot
President, American Society
of Civil Engineers
345 East 47th Street
New York, New York 10017

Dear Mr. Poirot:

I am pleased to furnish the American Society of Civil Engineers (ASCE) a copy of the U. S. Army Corps of Engineers Engineer Manual, EM 1110-1-1905, Bearing Capacity of Soils. The Corps uses this manual as the basis for calculating the bearing capacity of soils under shallow and deep foundations for footings, other structures and embankments. The manual also describes site exploration requirements.

I understand that ASCE plans to publish this manual for public distribution. I believe this will benefit the civil engineering community by improving transfer of technology between the Corps and other engineering professionals.

Sincerely,

Arthur E. Williams
Lieutenant General, U. S. Army
Commanding

CHAPTER 1

INTRODUCTION

1-1. Purpose and Scope

This manual presents guidelines for calculation of the bearing capacity of soil under shallow and deep foundations supporting various types of structures and embankments. This information is generally applicable to foundation investigation and design conducted by Corps of Engineer agencies.

A. APPLICABILITY. Principles for evaluating bearing capacity presented in this manual are applicable to numerous types of structures such as buildings and houses, towers and storage tanks, fills, embankments and dams. These guidelines may be helpful in determining soils that will lead to bearing capacity failure or excessive settlements for given foundations and loads.

B. EVALUATION. Bearing capacity evaluation is presented in Table 1-1. Consideration should be given to obtaining the services and advice of specialists and consultants in foundation design where foundation conditions are unusual or critical or structures are economically significant.

1. Definitions, failure modes and factors that influence bearing capacity are given in Chapter 1.

2. Evaluation of bearing capacity can be complicated by environmental and soil conditions. Some of these non-load related design considerations are given in Chapter 2.

3. Laboratory and in situ methods of determining soil parameters required for analysis of bearing capacity are given in Chapter 3.

4. Analysis of the bearing capacity of shallow foundations is given in Chapter 4 and of deep foundations is given in Chapter 5.

C. LIMITATIONS. This manual presents estimates of obtaining the bearing capacity of shallow and deep foundations for certain soil and foundation conditions using well-established, approximate solutions of bearing capacity.

1. This manual excludes analysis of the bearing capacity of foundations in rock.

2. This manual excludes analysis of bearing capacity influenced by seismic forces.

3. Refer to EM 1110-2-1902, Stability of Earth and Rockfill Dams, for solution of the slope stability of embankments.

D. REFERENCES. Standard references pertaining to this manual are listed in Appendix A, References. Each reference is identified in the text by the designated Government publication number or performing agency. Additional reading materials are listed in Appendix B, Bibliography.

1-2. Definitions

A. BEARING CAPACITY. Bearing capacity is the ability of soil to safely carry the pressure placed on the soil from any engineered structure without undergoing a shear failure with accompanying large settlements. Applying a bearing pressure which is safe with respect to failure does not ensure that settlement of the foundation will be within acceptable limits. Therefore, settlement analysis should generally be performed since most structures are sensitive to excessive settlement.

1. Ultimate Bearing Capacity. The generally accepted method of bearing capacity analysis is to assume that the soil below the foundation along a critical plane of failure (slip path) is on the verge of failure and to calculate the bearing pressure applied by the foundation required to cause this failure condition. This is the ultimate bearing capacity q_u. The general equation is

$$q_u = cN_c\zeta_c + \frac{1}{2}B\gamma'_H N_\gamma \zeta_\gamma + \sigma'_D N_q \zeta_q \qquad (1\text{-}1a)$$

$$Q_u = q_u BW \qquad (1\text{-}1b)$$

where

q_u = ultimate bearing capacity pressure, kips per square foot (ksf)

Q_u = ultimate bearing capacity force, kips

c = soil cohesion (or undrained shear strength C_u), ksf

B = foundation width, ft

W = foundation lateral length, ft

γ'_H = effective unit weight beneath foundation base within failure zone, kips/ft^3

σ'_D = effective soil or surcharge pressure at the foundation depth D, $\gamma'_D \cdot D$, ksf

γ'_D = effective unit weight of surcharge soil within depth D, kips/ft^3

N_c, N_γ, N_q = dimensionless bearing capacity factors for cohesion c, soil weight in the failure wedge, and surcharge q terms

$\zeta_c, \zeta_\gamma, \zeta_q$ = dimensionless correction factors for cohesion, soil weight in the failure wedge, and surcharge q terms accounting for foundation geometry and soil type

A description of factors that influence bearing capacity and calculation of γ'_H and γ'_D is given in Section 1-4. Details for calculation of the dimensionless bearing capacity N and correction ζ factors are given in Chapter 4 for shallow foundations and in Chapter 5 for deep foundations.

a. Bearing pressures exceeding the limiting shear resistance of the soil cause collapse of the structure which is usually accompanied by tilting. A bearing capacity failure results in very large downward movements of the structure, typically 0.5 ft to over 10 ft in magnitude. A bearing capacity failure of this type usually occurs within 1 day after the first full load is applied to the soil.

b. Ultimate shear failure is seldom a controlling factor in design because few structures are able to tolerate the rather large deformations that occur in soil prior to failure. Excessive settlement and differential movement can cause distortion and cracking in structures, loss of freeboard and water retaining capacity of embankments and dams, misalignment of operating equipment, discomfort to occupants, and eventually structural failure. Therefore, settlement analyses must frequently be performed to establish the expected foundation settlement. Both total and differential settlement between critical parts of the structure must be compared with allowable values. Refer to EM 1110-1-1904 for further details.

c.. Calculation of the bearing pressure required for ultimate shear failure is useful where sufficient data are not available to perform a settlement analysis. A suitable safety factor can be applied to the calculated ultimate bearing pressure where sufficient experience and practice have established appropriate safety factors. Structures such as embankments and uniformly loaded tanks, silos, and mats founded on soft soils and designed to tolerate large settlements all may be susceptible to a base shear failure.

Table 1-1. Bearing capacity evaluation

Step	Procedure
1	Evaluate the ultimate bearing capacity pressure q_u or bearing force Q_u using guidelines in this manual and Equation 1-1.
2	Determine a reasonable factor of safety *FS* based on available subsurface surface information, variability of the soil, soil layering and strengths, type and importance of the structure and past experience. *FS* will typically be between 2 and 4. Typical *FS* are given in Table 1-2.
3	Evaluate allowable bearing capacity q_a by dividing q_u by *FS*; i.e., $q_a = q_u/FS$, Equation 1-2a or $Q_a = Q_u/FS$, Equation 1-2b.
4	Perform settlement analysis when possible and adjust the bearing pressure until settlements are within tolerable limits. The resulting design bearing pressure q_d may be less than q_a. Settlement analysis is particularly needed when compressible layers are present beneath the depth of the zone of a potential bearing failure. Settlement analysis must be performed on important structures and those sensitive to settlement. Refer to EM 1110-1-1904 for settlement analysis of shallow foundations and embankments and EM 1110-2-2906, Reese and O'Neill (1988) and Vanikar (1986) for settlement of deep foundations.

2. Allowable Bearing Capacity. The allowable bearing capacity q_a is the ultimate bearing capacity q_u divided by an appropriate factor of safety FS,

$$q_d = \frac{q_u}{FS} \qquad (1\text{-}2a)$$

$$Q_d = \frac{Q_u}{FS} \qquad (1\text{-}2b)$$

FS is often determined to limit settlements to less than 1 inch and it is often in the range of 2 to 4.

a. Settlement analysis should be performed to determine the maximum vertical foundation pressures which will keep settlements within the predetermined safe value for the given structure. The recommended design bearing pressure q_d or design bearing

Table 1-2. Typical factors of safety

Structure	*FS*
Retaining	
Walls	3
Temporary braced excavations	>2
Bridges	
Railway	4
Highway	3.5
Buildings	
Silos	2.5
Warehouses	2.5*
Apartments, offices	3
Light industrial, public	3.5
Footings	3
Mats	>3
Deep Foundations	
With load tests	2
Driven piles with wave equation analysis calibrated to results of dynamic pile tests	2.5
Without load tests	3
Multilayer soils	4
Groups	3

*Modern warehouses often require superflat floors to accommodate modern transport equipment; these floors require extreme limitations to total and differential movements with *FS* > 3.

force Q_d could be less than q_a or Q_a due to settlement limitations.

b. When practical, vertical pressures applied to supporting foundation soils which are preconsolidated should be kept less than the maximum past pressure (preconsolidation load) applied to the soil. This avoids the higher rate of settlement per unit pressure that occurs on the virgin consolidation settlement portion of the e-log p curve past the preconsolidation pressure. The e-log p curve and preconsolidation pressure are determined by performing laboratory consolidation tests, EM 1110-2-1906.

3. Factors of Safety. Table 1-2 illustrates some factors of safety. These FS's are conservative and will generally limit settlement to acceptable values, but economy may be sacrificed in some cases.

a. *FS* selected for design depends on the extent of information available on subsoil characteristics and their variability. A thorough and extensive subsoil investigation may permit use of smaller *FS*.

b. *FS* should generally be ≥ 2.5 and never less than 2.

c. *FS* in Table 1-2 for deep foundations are consistent with usual compression loads. Refer to EM 1110-2-2906 for *FS* to be used with other loads.

B. SOIL. Soil is a mixture of irregularly shaped mineral particles of various sizes containing voids between particles. These voids may contain water if the soil is saturated, water and air if partly saturated, and air if dry. Under unusual conditions, such as sanitary landfills, gases other than air may be in the voids. The particles are a by-product of mechanical and chemical weathering of rock and described as gravels, sands, silts, and clays. Bearing capacity analysis requires a distinction between cohesive and cohesionless soils.

1. Cohesive Soil. Cohesive soils are fine-grained materials consisting of silts, clays, and/or organic material. These soils exhibit low to high strength when unconfined and when air-dried depending on specific characteristics. Most cohesive soils are relatively impermeable compared with cohesionless soils. Some silts may have bonding agents between particles such as soluble salts or clay aggregates. Wetting of soluble agents bonding silt particles may cause settlement.

2. Cohesionless Soil. Cohesionless soil is composed of granular or coarse-grained materials with visually detectable particle sizes and with little cohesion or adhesion between particles. These soils have little or no strength, particularly when dry, when unconfined and little or no cohesion when submerged. Strength occurs from internal friction when the material is confined. Apparent adhesion between particles in cohesionless soil may occur from capillary tension in the pore water. Cohesionless soils are usually relatively free-draining compared with cohesive soils.

C. FOUNDATIONS. Foundations may be classified in terms of shallow and deep elements and retaining structures that distribute loads from structures to the underlying soil. Foundations must be designed to maintain soil pressures at all depths within the allowable bearing capacity of the soil and also must limit total and differential movements to within levels that can be tolerated by the structure.

1. Shallow Foundations. Shallow foundations are usually placed within a depth D beneath the ground surface less than the minimum width *B* of the foundation. Shallow foundations consist of spread and continuous footings, wall footings and mats (Figure 1-1).

a. A spread footing distributes column or other loads from the structure to the soil, Figure 1-1a,

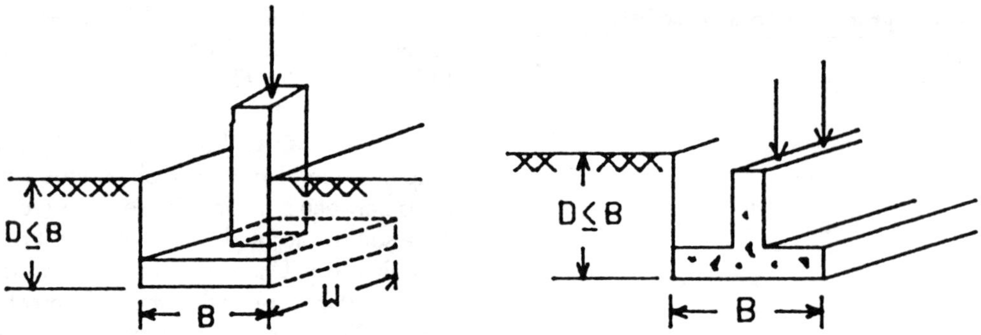

a. SPREAD FOOTING b. WALL FOOTING

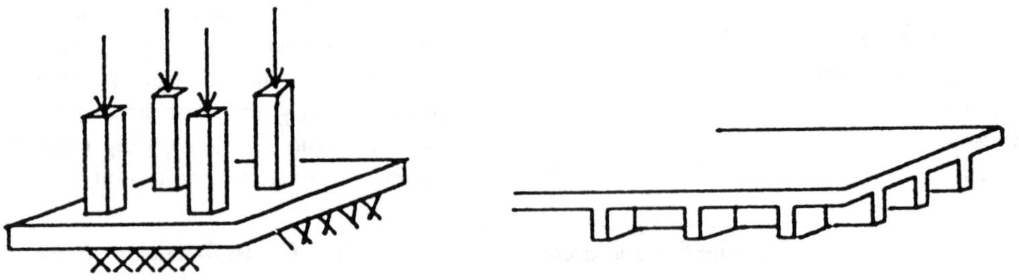

c. FLAT MAT WITH MULTIPLE COLUMNS d. RIBBED MAT

Figure 1-1. Shallow Foundations

where $B \leq W \leq 10B$. A continuous footing is a spread footing where $W > 10B$.

b. A wall footing is a long load bearing footing, Figure 1-1b.

c. A mat is continuous in two directions capable of supporting multiple columns, wall or floor loads. It has dimensions from 20 to 80 ft or more for houses and hundreds of feet for large structures such as multi-story hospitals and some warehouses, Figure 1-1c. Ribbed mats, Figure 1-1d, consisting of stiffening beams placed below a flat slab are useful in unstable soils such as expansive, collapsible or soft materials where differential movements can be significant (exceeding 0.5 inch).

2. Deep Foundations. Deep foundations can be as short as 15 to 20 ft or as long as 200 ft or more and may consist of driven piles, drilled shafts or stone columns (Figure 1-2). A single drilled shaft often has greater load bearing capacity than a single pile. Deep foundations may be designed to carry superstructure loads through poor soil (loose sands, soft clays, and collapsible materials) into competent bearing materials. Even when piles or drilled shafts are carried into competent materials, significant settlement can still occur if compressible soils are located below the tip of these deep foundations. Deep foundation support is usually more economical for depths less than 100 ft than mat foundations.

a. A pile may consist of a timber pole, steel pipe section, H-beam, solid or hollow precast concrete section or other slender element driven into the ground using pile driving equipment, Figure 1-2a. Pile foundations are usually placed in groups often with spacings S of 3 to $3.5B$ where B is the pile diameter. Smaller spacings are often not desirable because of the potential for pile intersection and a reduction in load carrying capacity . A pile cap is necessary to spread vertical and horizontal loads and any overturning moments to all of the piles in the group. The cap of onshore structures usually consists of reinforced con-

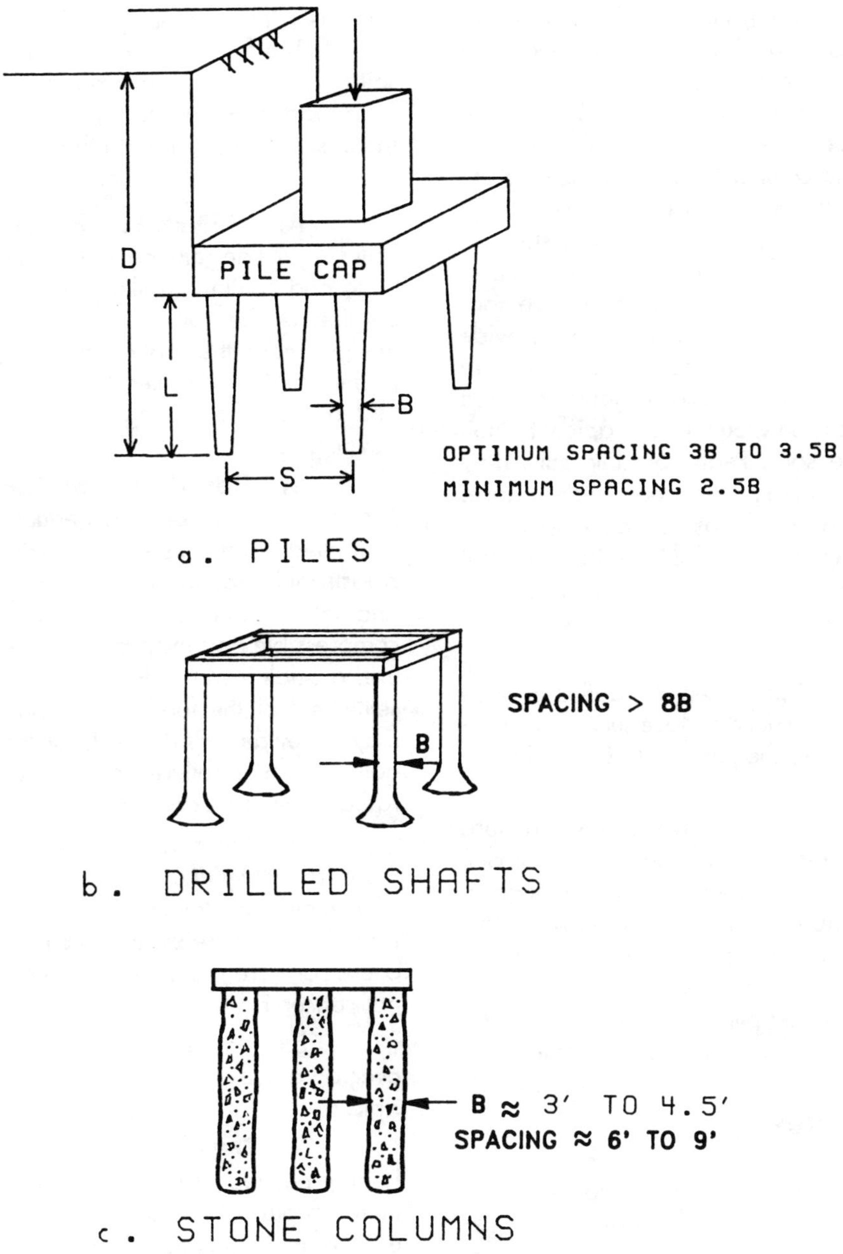

Figure 1-2. Deep foundations

crete cast on the ground, unless the soil is expansive. Offshore caps are often fabricated from steel.

b. A drilled shaft is a bored hole carried down to a good bearing stratum and filled with concrete, Figure 1-2b. A drilled shaft often contains a cage of reinforcement steel to provide bending, tension, and compression resistance. Reinforcing steel is always needed if the shaft is subject to lateral or tensile loading. Drilled shaft foundations are often placed as single elements beneath a column with spacings greater than 8 times the width or diameter of the shaft. Other names for drilled shafts include bored and underreamed pile, pier and caisson. Auger-cast or auger-grout piles are included in this category because these are not driven, but installed by advancing a continous-flight hollow-stem auger to the required depth and filling the hole created by the auger with grout under pressure as the auger is withdrawn. Diameters may vary from 0.5 to 10 ft or more. Spacings > 8*B* lead to minimal interaction between adjacent drilled shafts so that bearing capacity of these foundations may be analyzed using equations for single shafts. Shafts bearing in rock (rock drilled piers) are often placed closer than 8 diameters.

c. A stone column, Figure 1-2c, consists of granular (cohesionless) material of stone or sand often placed by vibroflotation in weak or soft subsurface soils with shear strengths from 0.2 to 1 ksf. The base of the column should rest on a dense stratum with adequate bearing capacity. The column is made by sinking the vibroflot or probe into the soil to the required depth using a water jet. While adding additional stone to backfill the cavity, the probe is raised and lowered to form a dense column. Stone columns usually are constructed to strengthen an area rather than to provide support for a limited size such as a single footing. Care is required when sensitive or peaty, organic soils are encountered. Construction should occur rapidly to limit vibration in sensitive soils. Peaty, organic soils may cause construction problems or poor performance. Stone columns are usually not as economical as piles or piers for supporting conventional type structures, but are competitive when used to support embankments on soft soils, slopes, and remedial or new work for preventing liquefaction.

d. The length L of a deep foundation may be placed at depths below ground surface such as for supporting basements where the pile length $L \leq D$, Figure 1-2a.

3. Retaining Structures. Any structure used to retain soil or other material in a shape or distribution different from that under the influence of gravity is a retaining structure. These structures may be permanent or temporary and consist of a variety of materials such as plain or reinforced concrete, reinforced soil, closely spaced piles or drilled shafts, and interlocking elments of wood, metal or concrete.

1-3. Failure Modes

The modes of potential failure caused by a footing of width B subject to a uniform pressure q develop the limiting soil shear strength τ_s at a given point along a slip path such as in Figure 1-3a

$$\tau_s = c + \sigma_n \tan \phi \tag{1-3}$$

where

τ_s = soil shear strength, ksf

c = unit soil cohesion (undrained shear strength C_u), ksf

σ_n = normal stress on slip path, ksf

ϕ = friction angle of soil, deg

From Figure 1-3a, the force on a unit width of footing causing shear is q_u times B, $q_u \cdot B$. The force resisting shear is τ_s times the length of the slip path '*abc*' or $\tau_s \cdot$ '*abc*'. The force resisting shear in a purely cohesive soil is $c \cdot$ '*abc*' and in a purely friction soil $\sigma_n \tan \phi \cdot$ '*abc*'. The length of the slip path '*abc*' resisting failure increases in proportion to the width of footing B.

A. GENERAL SHEAR. Figure 1-3a illustrates right side rotation shear failure along a well defined and continuous slip path '*abc*' that will result in bulging of the soil adjacent to the foundation. The wedge under the footing goes down and the soil is pushed to the side laterally and up. Surcharge above and outside the footing helps hold the block of soil down.

1. Description of Failure. Most bearing capacity failures occur in general shear under stress controlled conditions and lead to tilting and sudden catastrophic type movement. For example, dense sands and saturated clays loaded rapidly are practically incompressible and may fail in general shear. After failure, a small increase in stress causes large additional settlement of the footing. The bulging of surface soil may be evident on the side of the foundation undergoing a shear failure. In relatively rare cases, some radial tension cracks may be present.

a. Shear failure has been found to occur more frequently under shallow foundations supporting silos, tanks, and towers than under conventional buildings. Shear failure usually occurs on only one side because soils are not homogeneous and the load is often not concentric.

b. Figure 1-3b illustrates shear failure in soft over rigid soil. The failure surface is squeezed by the rigid soil.

2. Depth of Failure. Depth of shear zone H may be approximated by assuming that the maximum depth of shear failure occurs beneath the edge of the foundation, Figure 1-3a. If $\psi = 45 + \phi'/2$ (Vesic 1973), then

$$H = B \cdot \tan \psi \tag{1-4a}$$

$$H = B \cdot \tan\left(45 + \frac{\phi'}{2}\right) \tag{1-4b}$$

where

H = depth of shear failure beneath foundation base, ft

B = footing width, ft

$\psi = 45 + \phi'/2$, deg

ϕ' = effective angle of internal friction, deg

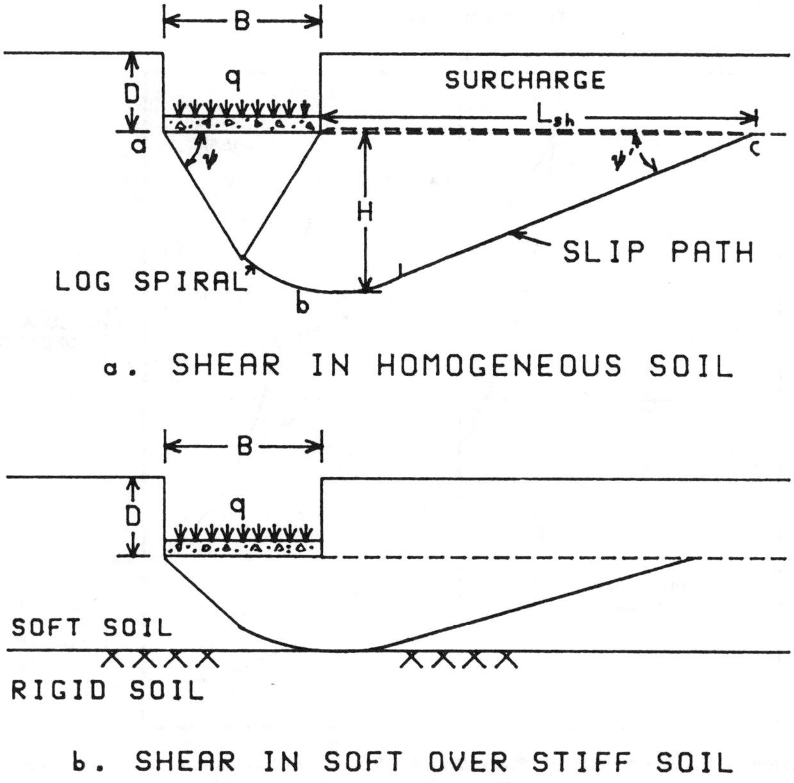

Figure 1-3. General shear failure

The depth H for a shear failure will be $1.73B$ if $\phi' = 30°$, a reasonable assumption for soils. H therefore should not usually be greater than $2B$. If rigid material lies within $2B$, then H will be $< 2B$ and will not extend deeper than the depth of rigid material, Figure 1-3b. Refer to Leonards (1962) for an alternative method of determining the depth of failure.

3. Horizontal Length of Failure. The length that the failure zone extends from the foundation perimeter at the foundation depth L_{sh}, Figure 1-3a, may be approximated by

$$L_{sh} = (H + D)\cot \psi' = (H + D)\tan \psi \qquad (1\text{-}5a)$$

$$L_{sh} = (H + D)\tan\left(45 + \frac{\phi'}{2}\right) \qquad (1\text{-}5b)$$

where D is the depth of the foundation base beneath the ground surface and $\psi' = 45 - \phi'/2$. L_{sh} $1.73(H + D)$ if $\phi' = 30°$. The shear zone may extend horizontally about $3B$ from the foundation base. Refer to Leonards (1962) for an alternative method of determining the length of failure.

B. PUNCHING SHEAR. Figure 1-4 illustrates punching shear failure along a wedge slip path '*abc*'. Slip lines do not develop and little or no bulging occurs at the ground surface. Vertical movement associated with increased loads causes compression of the soil immediately beneath the foundation. Figure 1-4 also illustrates punching shear of stiff over soft soil.

1. Vertical settlement may occur suddenly as a series of small movements without visible collapse or significant tilting. Punching failure is often associated with deep foundation elements, particularly in loose sands.

2. Local shear is a punching-type failure and it is more likely to occur in loose sands, silty sands, and weak clays. Local shear failure is characterized by a slip path that is not well defined except immediately

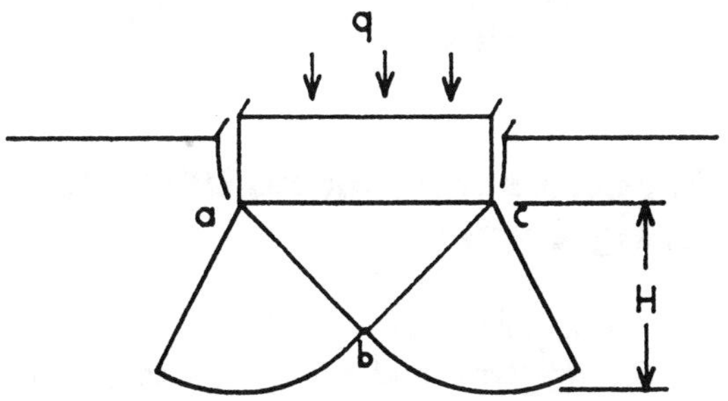

Figure 1-4. Punching failure

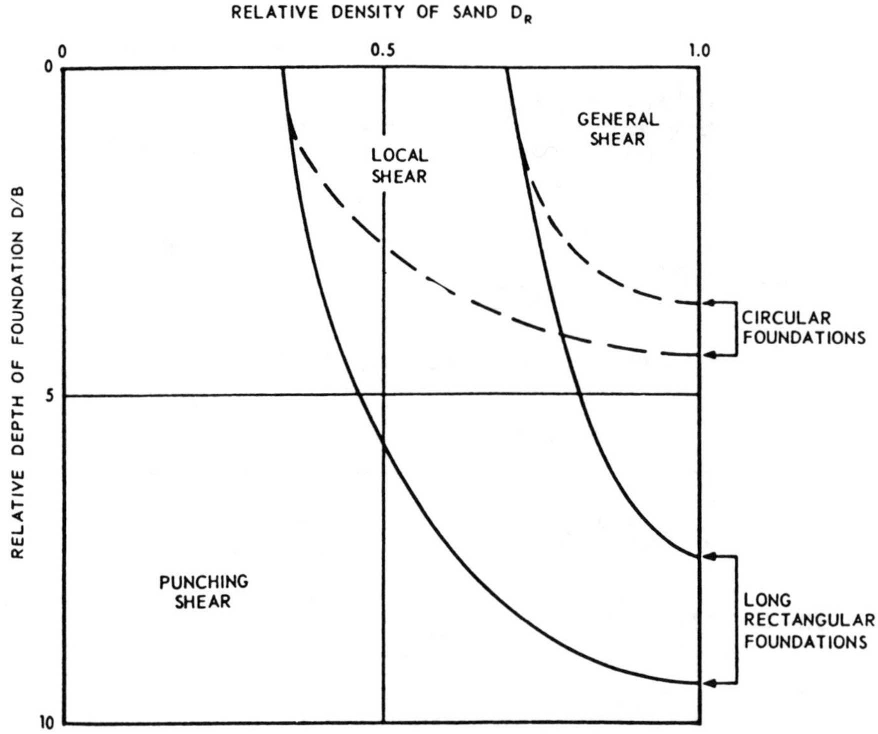

Figure 1-5. Variation of the nature of bearing capacity failure in sand with relative density D_R and relative depth D/B (Vesic 1963). Reprinted by permission of the Transportation Research Board, Highway Research Record 39, "Bearing Capacity of Deep Foundations in Sands" by A. B. Vesic, p. 136.

beneath the foundation. Failure is not catastrophic and tilting may be insignificant. Applied loads can continue to increase on the foundation soil following local shear failure.

C. FAILURE IN SAND. The approximate limits of types of failure to be expected at relative depths D/B and relative density of sand D_R vary as shown in Figure 1-5. There is a critical relative depth below which only punching shear failure occurs. For circular foundations, this critical relative depth is about $D/B = 4$ and for long ($L \approx 5B$) rectangular foundations around $D/B = 8$. The limits of the types of failure depend upon the compressibility of the sand. More compressible materials will have lower critical depths (Vesic 1963).

1-4. Factors Influencing Ultimate Bearing Capacity

Principal factors that influence ultimate bearing capacities are type and strength of soil, foundation width and depth, soil weight in the shear zone, and surcharge. Structural rigidity and the contact stress distribution do not greatly influence bearing capacity. Bearing capacity analysis assumes a uniform contact pressure between the foundation and underlying soil.

A. SOIL STRENGTH. Many sedimentary soil deposits have an inherent anisotropic structure due to their common natural deposition in horizontal layers. Other soil deposits such as saprolites may also exhibit anisotropic properties. The undrained strength of cohesive soil and friction angle of cohesionless soil will be influenced by the direction of the major principal stress relative to the direction of deposition. This manual calculates bearing capacity using strength parameters determined when the major principal stress is applied in the direction of deposition.

1. Cohesive Soil. Bearing capacity of cohesive soil is proportional to soil cohesion c if the effective friction angle ϕ' is zero.

2. Cohesionless Soil. Bearing capacity of cohesionless soil and mixed "c-ϕ" soils increases nonlinearly with increases in the effective friction angle.

B. FOUNDATION WIDTH. Foundation width influences ultimate bearing capacity in cohesion-

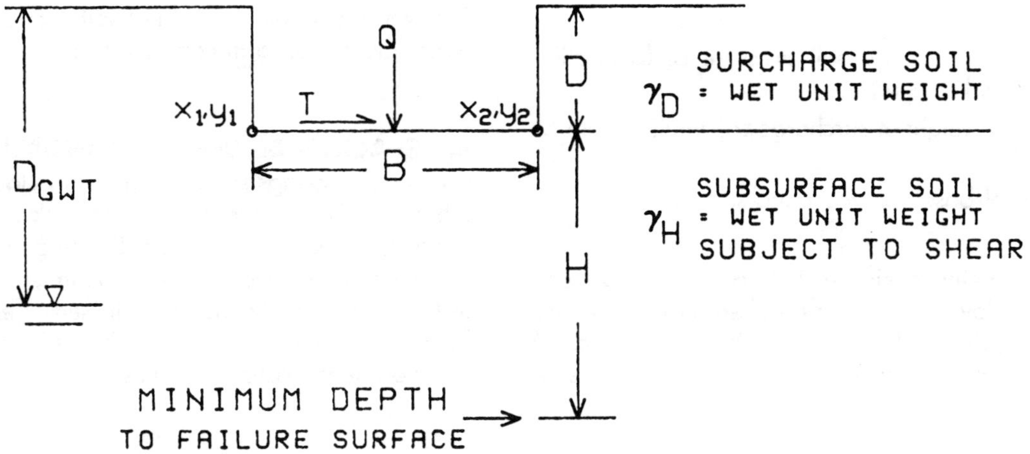

Figure 1-6. Schematic of foundation system

less soil. Foundation width also influences settlement, which is important in determining design loads. The theory of elasticity shows that, for an ideal soil whose properties do not change with stress level, settlement is proportional to foundation width.

1. Cohesive Soil. The ultimate bearing capacity of cohesive soil of infinite depth and constant shear strength is independent of foundation width because $c \cdot 'abc'/B$, Figure 1-3a, is constant.

2. Cohesionless Soil. The ultimate bearing capacity of a footing placed at the surface of a cohesionless soil where soil shear strength largely depends on internal friction is directly proportional to the width of the bearing area.

C. FOUNDATION DEPTH. Bearing capacity, particularly that of cohesionless soil, increases with foundation depth if the soil is uniform. Bearing capacity is reduced if the foundation is carried down to a weak stratum.

1. The bearing capacity of larger footings with a slip path that intersects a rigid stratum will be greater than that of a smaller footing with a slip path that does not intersect a deeper rigid stratum, Figure 1-3.

2. Foundations placed at depths where the structural weight equals the weight of displaced soil usually assures adequate bearing capacity and only recompression settlement. Exceptions include structures supported by underconsolidated soil and collapsible soil subject to wetting.

D. SOIL WEIGHT AND SURCHARGE. Subsurface and surcharge soil weights contribute to bearing capacity as given in Equation 1-1. The depth to the water table influences the subsurface and surcharge soil weights (Figure 1-6). Water table depth can vary significantly with time.

1. If the water table is below the depth of the failure surface, then the water table has no influence on the bearing capacity and effective unit weights γ'_D and γ'_H in Equation 1-1 are equal to the wet unit weight of the soils γ_D and γ_H.

2. If the water table is above the failure surface and beneath the foundation base, then the effective unit weight γ'_H can be estimated as

$$\gamma'_H = \gamma_{HSUB} + \frac{D_{GWT} - D}{H} \cdot \gamma_w \qquad (1\text{-}6)$$

where

γ_{HSUB} = submerged unit weight of subsurface soil, $\gamma_H - \gamma_w$, kips/ft^3

D_{GWT} = depth below ground surface to groundwater, ft

H = minimum depth below base of foundation to failure surface, ft

γ_w = unit weight of water, 0.0625 kip/ft^3

3. The water table should not be above the base of the foundation to avoid construction, seepage, and uplift problems. If the water table is above the base of the foundation, then the effective surcharge term σ'_D may be estimated by

$$\sigma'_D = \gamma'_D \cdot D \qquad (1\text{-}7a)$$

$$\gamma'_D = \gamma_D - \frac{D - D_{GWT}}{D} \cdot \gamma_w \qquad (1\text{-}7b)$$

where

σ'_D = effective surcharge soil pressure at foundation depth D, ksf

γ_D = unit wet weight of surcharge soil within depth D, kips/ft^3

D = depth of base below ground surface, ft

4. Refer to Figure 2, Chapter 4 in Department of the Navy (1982), for an alternative procedure of estimating depth of failure zone H and influence of groundwater on bearing capacity in cohesionless soil. The wet or saturated weight of soil above or below the water table is used in cohesive soil.

E. SPACING BETWEEN FOUNDATIONS. Foundations on footings spaced sufficiently close together to intersect adjacent shear zones may decrease bearing capacity of each foundation. Spacings between footings should be at least 1.5B, to minimize any reduction in bearing capacity. Increases in settlement of existing facilities should be checked when placing new construction near existing facilities.

CHAPTER 2

NON-LOAD RELATED DESIGN CONSIDERATIONS

2-1. General

Special circumstances may complicate the evaluation of bearing capacity such as earthquake and dynamic motion, soil subject to frost action, subsurface voids, effects of expansive and collapsible soil, earth reinforcement, heave in cuts and scour, and seepage erosion. This chapter briefly describes these applications. Coping with soil movements and ground improvement methods are discussed in TM 5-818-7, EM 1110-1-1904 and EM 1110-2-3506.

2-2. Earthquake and Dynamic Motion

Cyclic or repeated motion caused by seismic forces or earthquakes, vibrating machinery, and other disturbances such as vehicular traffic, blasting and pile driving may cause pore pressures to increase in foundation soil. As a result, bearing capacity will be reduced from the decreased soil strength. The foundation soil can liquify when pore pressures equal or exceed the soil confining stress reducing effective stress to zero and causes gross differential settlement of structures and loss of bearing capacity. Structures supported by shallow foundations can tilt and exhibit large differential movement and structural damage. Deep foundations lose lateral support as a result of liquefaction and horizontal shear forces lead to buckling and failure. The potential for soil liquefaction and structural damage may be reduced by various soil improvement methods.

A. CORPS OF ENGINEER METHOD. Methods of estimating bearing capacity of soil subject to dynamic action depend on methods of correcting for the change in soil shear strength caused by changes in pore pressure. Differential movements increase with increasing vibration and can cause substantial damage to structures. Department of the Navy (1983), "Soil Dynamics, Deep Stabilization, and Special Geotechnical Construction", describes evaluation of vibration induced settlement.

B. COHESIVE SOIL. Dynamic forces on conservatively designed foundations with $FS \geq 3$ will probably have little influence on performance of structures. Limited data indicate that strength reduction during cyclic loading will likely not exceed 20 percent in medium to stiff clays (Edinger 1989). However, vibration induced settlement should be estimated to be sure structural damages will not be significant.

C. COHESIONLESS SOIL. Dynamic forces may significantly reduce bearing capacity in sand. Foundations conservatively designed to support static and earthquake forces will likely fail only during severe earthquakes and only when liquefaction occurs (Edinger 1989). Potential for settlement large enough to adversely influence foundation performance is most likely in deep beds of loose dry sand or saturated sand subject to liquefaction. Displacements leading to structural damage can occur in more compact sands, even with relative densities approaching 90 percent, if vibrations are sufficient. The potential for liquefaction should be analyzed as described in EM 1110-1-1904.

2-3. Frost Action

Frost heave in certain soils in contact with water and subject to freezing temperatures or loss of strength of frozen soil upon thawing can alter bearing capacity over time. Frost heave at below freezing temperatures occurs from formation of ice lenses in frost susceptible soil. As water freezes to increase the volume of the ice lense, the pore pressure of the remaining unfrozen water decreases and tends to draw additional warmer water from deeper depths. Soil below the depth of frost action tends to become dryer and consolidate, while soil within the depth of frost action tends to be wetter and contain fissures. The base of foundations should be below the depth of frost action. Refer to TM 5-852-4 and Lobacz (1986).

A. FROST SUSCEPTIBLE SOILS. Soils most susceptible to frost action are low cohesion ma-

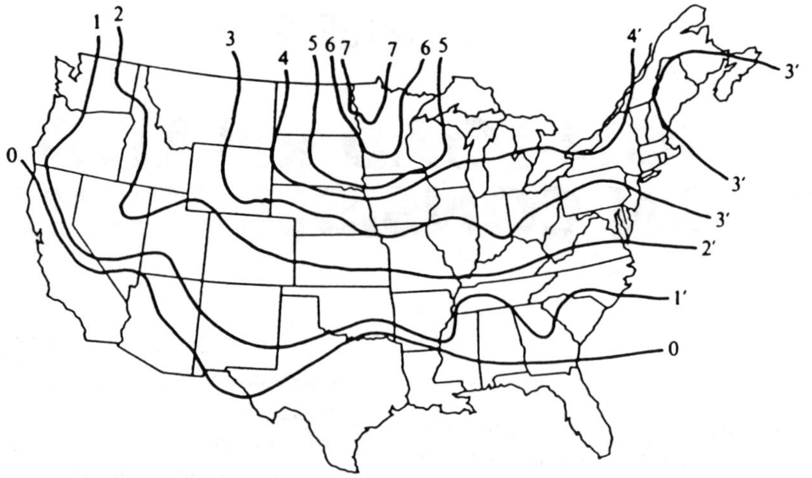

Figure 2-1. Approximate frost-depth contours in the United States. Reprinted by permission of McGraw-Hill Book Company, "Foundation Analysis and Design", p. 305, 1988, by J. E. Bowles.

terials containing a high percentage of silt-sized particles. These soils have a network of pores and fissures that promote migration of water to the freezing front. Common frost susceptible soils include silts (ML, MH), silty sands (SM), and low plasticity clays (CL, CL-ML).

B. DEPTH OF FROST ACTION. The depth of frost action depends on the air temperature below freezing and duration, surface cover, soil thermal conductivity and permeability and soil water content. Refer to TM 5-852-4 for methodology to estimate the depth of frost action in the United States from air-freezing index values. TM 5-852-6 provides calculation methods for determining freeze and thaw depths in soils. Figure 2-1 provides approximate frost-depth contours in the United States.

C. CONTROL OF FROST ACTION. Methods to reduce frost action are preferred if the depth and amount of frost heave is unpredictable.

1. Replace frost-susceptible soils with materials unaffected by frost such as clean medium to coarse sands and clean gravels, if these are readily available.

2. Pressure inject the soil with lime slurry or lime-flyash slurry to decrease the mass permeability.

3. Restrict the groundwater supply by increased drainage and/or an impervious layer of asphalt, plastic or clay.

4. Place a layer of thermal insulation such as foamed plastic or glass.

2-4. Subsurface Voids

A subsurface void influences and decreases bearing capacity when located within a critical depth D_c beneath the foundation. The critical depth is that depth below which the influence of pressure in the soil from the foundation is negligible. Evaluation of D_c is described in Section 3-3b.

A. VOIDS. Voids located beneath strip foundations at depth ratios $D_c/B > 4$ cause little influence on bearing capacity for strip footings. B is the foundation width. The critical depth ratio for square footings is about 2.

B. BEARING CAPACITY. The bearing capacity of a strip footing underlain by a centrally located void at ratios $D_c/B < 4$ decreases with increasing load eccentricity similar to that for footings without voids, but the void reduces the effect of load eccentricity. Although voids may not influence bearing capacity initially, these voids can gradually migrate upward with time in karst regions.

C. COMPLICATIONS OF CALCULATION. Load eccentricity and load inclination complicate calculation of bearing capacity when the void is close to the footing. Refer to Wang, Yoo and Hsieh (1987) for further information.

2-5. Expansive and Collapsible Soils

These soils change volume from changes in water content leading to total and differential foundation movements. Seasonal wetting and drying cycles have caused soil movements that often lead to excessive long-term deterioration of structures with substantial accumulative damage. These soils can have large strengths and bearing capacity when relatively dry.

A. EXPANSIVE SOIL. Expansive soils consist of plastic clays and clay shales that often contain colloidal clay minerals such as the montmorillonites. They include marls, clayey siltstone and sandstone, and saprolites. Some of these soils, especially dry residual clayey soil, may heave under low applied pressure but collapse under higher pressure. Other soils may collapse initially but heave later on. Estimates of the potential heave of these soils are necessary for consideration in the foundation design.

1. Identification. Degrees of expansive potential may be indicated as follows (Snethen, Johnson, and Patrick 1977):

Degree of Expansion	Liquid Limit, %	Plasticity Index, %	Natural Soil Suction, tsf
High	> 60	> 35	> 4.0
Marginal	50–60	25–35	1.5–4.0
Low	< 50	< 25	< 1.5

Soils with Liquid Limit (LL) < 35 and Plasticity Index (PI) < 12 have no potential for swell and need not be tested.

2. Potential Heave. The potential heave of expansive soils should be determined from results of consolidometer tests, ASTM D 4546. These heave estimates should then be considered in determining preparation of foundation soils to reduce destructive differential movements and to provide a foundation of sufficient capacity to withstand or isolate the expected soil heave. Refer to TM 5-818-7 and EM 1110-1-1904 for further information on analysis and design of foundations on expansive soils.

B. COLLAPSIBLE SOIL. Collapsible soils will settle without any additional applied pressure when sufficient water becomes available to the soil. Water weakens or destroys bonding material between particles that can severely reduce the bearing capacity of the original soil. The collapse potential of these soils must be determined for consideration in the foundation design.

1. Identification. Many collapsible soils are mudflow or windblown silt deposits of loess often found in arid or semiarid climates such as deserts, but dry climates are not necessary for collapsible soil. Typical collapsible soils are lightly colored, low in plasticity with LL < 45, PI < 25 and with relatively low densities between 65 and 105 lbs/ft^3 (60 to 40 percent porosity). Collapse rarely occurs in soil with porosity less than 40 percent. Refer to EM 1110-1-1904 for methods of identifying collapsible soil.

2. Potential Collapse. The potential for collapse should be determined from results of a consolidometer test as described in EM 1110-1-1904. The soil may then be modified as needed using soil improvement methods to reduce or eliminate the potential for collapse.

2-6. Soil Reinforcement

Soil reinforcement allows new construction to be placed in soils that were originally less than satisfactory. The bearing capacity of weak or soft soil may be substantially increased by placing various forms of reinforcement in the soil such as metal ties, strips, or grids, geotextile fabrics, or granular materials.

A. EARTH REINFORCEMENT. Earth reinforcement consists of a bed of granular soil strengthened with horizontal layers of flat metal strips, ties, or grids of high tensile strength material that develop a good frictional bond with the soil. The bed of reinforced soil must intersect the expected slip paths of shear failure, Figure 1-3a. The increase in bearing capacity is a function of the tensile load developed in any tie, breaking strength and pullout friction resistance of each tie and the stiffness of the soil and reinforcement materials.

1. An example calculation of the design of a reinforced slab is provided in Binquet and Lee (1975).

2. Slope stability package UTEXAS2 (Edris 1987) may be used to perform an analysis of the bearing capacity of either the unreinforced or reinforced soil beneath a foundation. A small slope of about 1 degree must be used to allow the computer program to operate. The program will calculate the bearing capacity of the weakest slip path, Figure 1-3a, of infinite length (wall) footings, foundations, or embankments.

B. GEOTEXTILE HORIZONTAL REINFORCEMENT. High-strength geotextile fabrics placed on the surface under the proper conditions allow construction of embankments and other structures on soft foundation soils that normally will not otherwise support pedestrian traffic, vehicles, or conventional construction equipment. Without adequate soil reinforcement, the embankment may fail during or after construction by shallow or deep-seated sliding wedge or circular arc-type failures or by excessive subsidence caused by soil creep, consolidation, or bearing capacity shear failure. Fabrics can contribute toward a solution to these problems. Refer to TM 5-800-08 for further information on analysis and design of embankment slope stability, embankment sliding, embankment spreading, embankment rotational displacement, and longitudinal fabric strength reinforcement.

1. Control of Horizontal Spreading. Excessive horizontal sliding, splitting, and spreading of embankments and foundation soils may occur from large lateral earth pressures caused by embankment soils. Fabric reinforcement between a soft foundation soil and embankment fill materials provides forces that resist the tendency to spread horizontally. Failure of fabric-reinforced embankments may occur by slippage between the fabric and fill material, fabric tensile failure, or excessive fabric elongation. These failure modes may be prevented by specifyng fabrics of required soil-fabric friction, tensile strength, and tensile modulus.

2. Control of Rotational Failure. Rotational slope and/or foundation failures are resisted by the use of fabrics with adequate tensile strength and embankment materials with adequate shear strength. Rotational failure occurs through the embankment, foundation layer, and the fabric. The tensile strength of the fabric must be sufficiently high to control the large unbalanced rotational moments. Computer program UTEXAS2 (Edris 1987) may be used to determine slope stability analysis with and without reinforcement to aid in the analysis and design of embankments on soft soil.

3. Control of Bearing Capacity Failure. Soft foundations supporting embankments may fail in bearing capacity during or soon after construction before consolidation of the foundation soil can occur. When consolidation does occur, settlement will be similar for either fabric reinforced or unreinforced embankments. Settlement of fabric reinforced embankments will often be more uniform than non-reinforced embankments.

a. Fabric reinforcement helps to hold the embankment together while the foundation strength increases through consolidation.

b. Large movements or center sag of embankments may be caused by improper construction such as working in the center of the embankment before the fabric edges are covered with fill material to provide a berm and fabric anchorage. Fabric tensile strength will not be mobilized and benefit will not be gained from the fabric if the fabric is not anchored.

c. A bearing failure and center sag may occur when fabrics with insufficient tensile strength and modulus are used, when steep embankments are constructed, or when edge anchorage of fabrics is insufficient to control embankment splitting. If the bearing capacity of the foundation soil is exceeded, the fabric must elongate to develop the required fabric stress to support the embankment load. The foundation soil will deform until the foundation is capable of carrying the excessive stresses that are not carried in the fabric. Complete failure occurs if the fabric breaks.

C. GRANULAR COLUMN IN WEAK SOIL.

A granular column supporting a shallow rectangular footing in loose sand or weak clay will increase the ultimate bearing capacity of the foundation.

1. The maximum bearing capacity of the improved foundation of a granular column supporting a rectangular foundation of identical cross-section is given approximately by (Das 1987)

$$q_u = K_p\left[\gamma_c D + 2\left(1 + \frac{B}{L}\right)C_u\right] \qquad (2\text{-}1)$$

where

K_p = Rankine passive pressure coefficient, $\dfrac{1 + \sin \phi_g}{1 - \sin \phi_g}$

ϕ_g = friction angle of granular material, degrees

γ_c = moist unit weight of weak clay, kip/ft^3

D = depth of the rectangle foundation below ground surface, ft

B = width of foundation, ft

L = length of foundation, ft

C_u = undrained shear strength of weak clay, ksf

Equation 2-1 is based on the assumption of a bulging failure of the granular column.

2. The minimum height of the granular column required to support the footing and to obtain the maximum increase in bearing capacity is 3*B*.

3. Refer to Bachus and Barksdale (1989) and Barksdale and Bachus (1983) for further details on analysis of bearing capacity of stone columns.

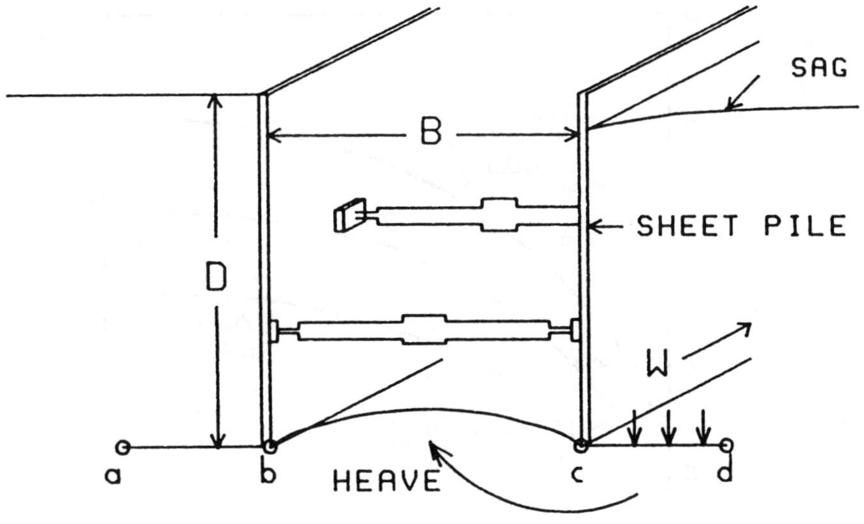

Figure 2-2. Heave failure in an excavation

2-7. Heaving Failure in Cuts

Open excavations in deep deposits of soft clay may fail by heaving because the weight of clay beside the cut pushes the underlying clay up into the cut (Figure 2-2 Terzaghi and Peck 1967). This results in a loss of ground at the ground surface. The bearing capacity of the clay at the bottom of the cut is $C_u N_c$. The bearing capacity factor N_c depends on the shape of the cut. N_c may be taken equal to that for a footing of the same B/W and D/B ratios as provided by the chart in Figure 2-3, where B is the excavation width, W is the excavation length, and D is the excavation depth below ground surface.

A. FACTOR OF SAFETY. *FS* against a heave failure should be at least 1.5. *FS* resisting heave at the excavation bottom caused by seepage should be 1.5 to 2.0 (TM 5-818-5).

$$FS = \frac{C_u N_c}{\gamma D} > 1.5 \qquad (2\text{-}2)$$

B. MINIMIZING HEAVE FAILURE. Extending continuous sheet pile beneath the bottom of the excavation will reduce the tendency for heave.

1. Sheet pile, even if the clay depth is large, will reduce flow into the excavation compared with pile and lagging support.

2. Driving the sheet pile into a hard under lying stratum below the excavation greatly reduces the tendency for a heave failure.

2-8. Soil Erosion and Seepage

Erosion of soil around and under foundations and seepage can reduce bearing capacity and can cause foundation failure.

A. SCOUR. Foundations such as drilled shafts and piles constructed in flowing water will cause the flow to divert around the foundation. The velocity of flow will increase around the foundation and can cause the flow to separate from the foundation. A wake develops behind the foundation and turbulence can occur. Eddy currents contrary to the stream flow is the basic scour mechanism. The foundation must be constructed at a sufficient depth beneath the maximum scour depth to provide sufficient bearing capacity.

1. Scour Around Drilled Shafts or Piles in Seawater. The scour depth may be estimated from empirical and experimental studies. Refer to Herbich, Schiller and Dunlap (1984) for further information.

a. The maximum scour depth to wave height ratio is ≤ 0.2 for a medium to fine sand.

b. The maximum depth of scour S_u as a function of Reynolds number R_θ is (Herbich, Schiller and Dunlap 1984)

$$S_u = 0.00073 R_\theta^{0.619} \qquad (2\text{-}3)$$

where S_u is in feet.

2. Scour Around Pipelines. Currents near pipelines strong enough to cause scour will grad-

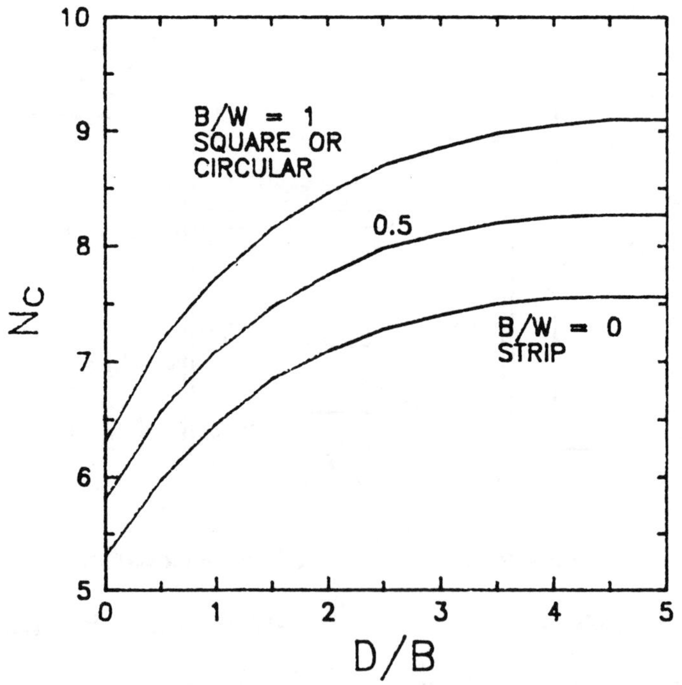

Figure 2-3. Estimation of bearing capacity factor N_c for heave in an excavation (Data from Terzaghi and Peck 1967)

ually erode away the soil causing the pipeline to lose support. The maximum scour hole depth may be estimated using methodology in Herbich, Schiller, and Dunlap (1984).

3. Mitigation of Scour. Rock-fill or riprap probably provides the easiest and most economical scour protection.

B. SEEPAGE. Considerable damage can occur to structures when hydrostatic uplift pressure beneath foundations and behind retaining walls becomes too large. The uplift pressure head is the height of the free water table when there is no seepage. If seepage occurs, flow nets may be used to estimate uplift pressure. Uplift pressures are subtracted from total soil pressure to evaluate effective stresses. Effective stresses should be used in all bearing capacity calculations.

1. Displacement piles penetrating into a confined hydrostatic head will be subject to uplift and may raise the piles from their end bearing.

2. Seepage around piles can reduce skin friction. Skin friction resistance can become less than the hydrostatic uplift pressure and can substantially reduce bearing capacity. Redriving piles or performing load tests after a waiting period following construction can determine if bearing capacity is sufficient.

CHAPTER 3

SOIL PARAMETERS

3-1. Methodology

A site investigation and soil exploration program of the proposed construction area should be initially completed to obtain data required for analysis of bearing capacity. Estimates of ultimate and allowable bearing capacity using analytical equations that model the shear failure of the structure along slip surfaces in the soil and methods for analyzing in situ test results that model the bearing pressures of the full size structure in the soil may then be carried out as described in Chapter 4 for shallow foundations and Chapter 5 for deep foundations. The scope of the analysis depends on the magnitude of the project and on how critical the bearing capacity of the soil is to the performance of the structure.

A. SOIL PARAMETERS. The natural variability of soil profiles requires realistic assessment of soil parameters by soil exploration and testing. Soil parameters required for analysis of bearing capacity are shear strength, depth to groundwater or the pore water pressure profile, and the distribution of total vertical overburden pressure with depth. The shear strength parameters required are the undrained shear strength C_u of cohesive soils, the effective angle of internal friction ϕ' for cohesionless soils, and the effective cohesion c' and angle of internal friction ϕ' for mixed soils that exhibit both cohesion and friction.

B. USE OF JUDGMENT. Judgment is required to characterize the foundation soils into one or a few layers with idealized parameters. The potential for long-term consolidation and settlement must be determined, especially where soft, compressible soil layers are present beneath the foundation. Assumptions made by the designer may significantly influence recommendations of the foundation design.

C. ACCEPTABILITY OF ANALYSIS. Acceptability of the bearing pressures applied to the foundation soil is usually judged by factors of safety applied to the ultimate bearing capacity and estimates made of potential settlement for the bearing pressures allowed on the foundation soil. The dimensions of the foundation or footing may subsequently be adjusted if required.

3-2. Site Investigation

Initially, the behavior of existing structures supported on similar soil in the same locality should be determined as well as the applied bearing pressures. These findings should be incorporated, using judgment, into the foundation design. A detailed subsurface exploration including disturbed and undisturbed samples for laboratory strength tests should then be carried out. Bearing capacity estimates may also be made from results of in situ soil tests. Refer to EM 1110-1-1804 for further information on site investigations.

A. EXAMINATION OF EXISTING RECORDS. A study of available service records and, where practical, a field inspection of structures supported by similar foundations in the bearing soil will furnish a valuable guide to probable bearing capacities.

1. Local Building Codes. Local building codes may give presumptive allowable bearing pressures based on past experience. This information should only be used to supplement the findings of in situ tests and analyses using one or more methods discussed subsequently because actual field conditions, and hence bearing pressures, are rarely identical with those conditions used to determine the presumptive allowable bearing pressures.

2. Soil Exploration Records. Existing records of previous site investigations near the proposed construction area should be examined to determine the general subsurface condition including the types of soils likely to be present, probable depths to groundwater level and changes in groundwater level, shear strength parameters, and compressibility characteristics.

B. SITE CHARACTERISTICS. The proposed construction site should be examined for plastic-

Table 3-1. Angle of internal friction of sands, ϕ'

a. Relative Density and Gradation
(Data from Schmertmann 1978)

Relative Density D_r, Percent	Fine-Grained Uniform	Fine-Grained Well-Graded	Medium-Grained Uniform	Medium-Grained Well-Graded	Coarse-Grained Uniform	Coarse-Grained Well-Graded
40	34	36	36	38	38	41
60	36	38	38	41	41	43
80	39	41	41	43	43	44
100	42	43	43	44	44	46

b. Relative Density and In Situ Soil Tests

Soil Type	Relative Density D_r, Percent	Standard Penetration Resistance N_{60} (Terzaghi and Peck 1967)	Cone Penetration Resistance q_c, ksf (Meyerhof 1974)	Friction Angle ϕ', deg: Meyerhof (1974)	Friction Angle ϕ', deg: Peck, Hanson and Thornburn (1974)	Friction Angle ϕ', deg: Meyerhof (1974)
Very Loose	< 20	< 4	—	< 30	< 29	< 30
Loose	20–40	4–10	0–100	30–35	29–30	30–35
Medium	40–60	10–30	100–300	35–38	30–36	35–40
Dense	60–80	30–50	300–500	38–41	36–41	40–45
Very Dense	> 80	> 50	500–800	41–44	> 41	> 45

ity and fissures of surface soils, type of vegetation, and drainage pattern.

1. Desiccation Cracking. Numerous desiccation cracks, fissures, and even slickensides can develop in plastic, expansive soils within the depth subject to seasonal moisture changes, the active zone depth Z_a, due to the volume change that occurs during repeated cycles of wetting and drying (desiccation). These volume changes can cause foundation movements that control the foundation design.

2. Vegetation. Vegetation desiccates the foundation soil from transpiration through leaves. Heavy vegetation such as trees and shrubs can desiccate foundation soil to substantial depths exceeding 50 or 60 ft. Removal of substantial vegetation in the proposed construction area may lead to significantly higher water tables after construction is complete and may influence bearing capacity.

3. Drainage. The ground surface should be sloped to provide adequate runoff of surface and rainwater from the construction area to promote trafficability and to minimize future changes in ground moisture and soil strength. Minimum slope should be 1 percent.

4. Performance of Adjacent Structures. Distortion and cracking patterns in nearby structures indicate soil deformation and the possible presence of expansive or collapsible soils.

C. IN SITU SOIL TESTS. In the absence of laboratory shear strength tests, soil strength parameters required for bearing capacity analysis may be estimated from results of in situ tests using empirical correlation factors. Empirical correlation factors should be verified by comparing estimated values with shear strengths determined from laboratory tests. The effective angle of internal friction ϕ' of cohesionless soil is frequently estimated from field test results because of difficulty in obtaining undisturbed cohesionless soil samples for laboratory soil tests.

1. Relative Density and Gradation. Relative density and gradation can be used to estimate the friction angle of cohesionless soils, Table 3-1a. Relative density is a measure of how dense a sand is compared with its maximum density.

a. ASTM D 653 defines relative density as the ratio of the difference in void ratio of a cohesionless soil in the loosest state at any given void ratio to the difference between the void ratios in the loosest and in the densest states. A very loose sand has a relative density of 0 percent and 100 percent in the densest

Table 3-2. Relative density and N_{60}

a. Rod Energy Correction Factor C_{ER}
(Data from Tokimatsu and Seed 1987)

Country	Hammer	Hammer Release	C_{ER}
Japan	Donut	Free-Fall	1.3
	Donut	Rope and Pulley with special throw release	1.12*
USA	Safety	Rope and Pulley	1.00*
	Donut	Rope and Pulley	0.75
Europe	Donut	Free-Fall	1.00*
China	Donut	Free-Fall	1.00*
	Donut	Rope and Pulley	0.83

*Methods used in USA.

b. Correction Factor C_N
(Data from Tokimatsu and Seed 1984)

C_N	σ'_{vo}*, ksf
1.6	0.6
1.3	1.0
1.0	2.0
0.7	4.0
0.55	6.0
0.50	8.0

*σ'_{vo} = effective overburden pressure.

c. Relative Density versus N_{60}
(Data from Jamiolkowski et al. 1988)

Sand	D_r, Percent	N_{60}
Very Loose	0– 15	0– 3
Loose	15– 35	3– 8
Medium	35– 65	8–25
Dense	65– 85	25–42
Very Dense	85–100	42–58

possible state. Extremely loose honeycombed sands may have a negative relative density.

b. Relative density may be calculated using standard test methods ASTM D 4254 and the void ratio of the in situ cohesionless soil,

$$D_r = \frac{e_{max} - e}{e_{max} - e_{min}} \cdot 100 \qquad (3\text{-}1a)$$

$$e = \frac{G}{\gamma_d}\gamma_w - 1 \qquad (3\text{-}1b)$$

where

e_{min} = reference void ratio of a soil at the maximum density

e_{max} = reference void ratio of a soil at the minimum density

G = specific gravity

γ_d = dry density, kips/ft^3

γ_w = unit weight of water, 0.0625 kip/ft^3

The specific gravity of the mineral solids may be determined using standard test method ASTM D 854. The dry density of soils that may be excavated can be determined in situ using standard test method ASTM D 1556.

2. Standard Penetration Test (SPT). The standard penetration resistance value N_{SPT}, often referred to as the blowcount, is frequently used to estimate the relative density of cohesionless soil. N_{SPT} is the number of blows required to drive a standard split-spoon sampler (1.42″ I.D., 2.00″ O.D.) 1 ft. The split-spoon sampler is driven by a 140-lb hammer falling 30 inches. The sampler is driven 18 inches and blows counted for the last 12 inches. N_{SPT} may be determined using standard method ASTM D 1586.

a. The N_{SPT} value may be normalized to an effective energy delivered to the drill rod at 60 percent of theoretical free-fall energy

$$N_{60} = C_{ER} \cdot C_N \cdot N_{SPT} \qquad (3\text{-}2)$$

where

N_{60} = penetration resistance normalized to an effective energy delivered to the drill rod at 60 percent of theoretical free-fall energy, blows/ft

C_{ER} = rod energy correction factor, Table 3-2a

C_N = overburden correction factor, Table 3-2b

N_{SPT} may have an effective energy delivered to the drill rod 50 to 55 percent of theoretical free fall energy.

b. Table 3-1 illustrates some relative density and N_{60} correlations with the angle of internal friction. Relative density may also be related with N_{60} through Table 3-2c.

c. The relative density of sands may be estimated from the N_{SPT} by (Data from Gibbs and Holtz 1957)

$$D_r \approx 100\left(\frac{N_{SPT}}{12\sigma'_{vo} + 17}\right)^{0.5} \qquad (3\text{-}3a)$$

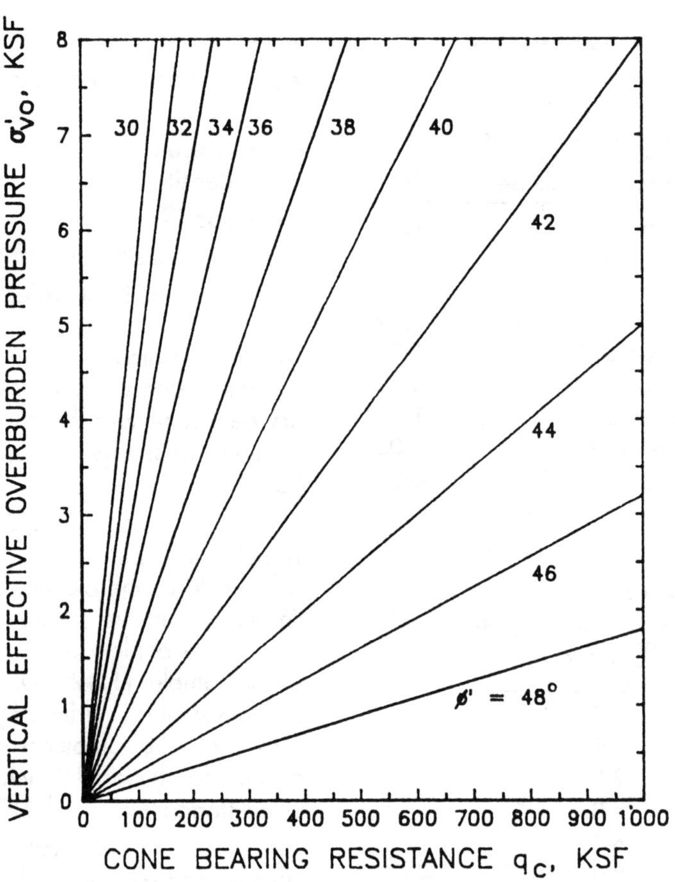

Figure 3-1. Approximate correlation between cone penetration resistance, peak effective friction angle and vertical effective overburden pressure for uncemented quartz sand (After Robertson and Campanella 1983)

where D_r is in percent and σ'_{vo} is the effective vertical overburden pressure, ksf.

d. The relative density of sands may also be estimated from N_{60} by (Jamiolkowski et al. 1988, Skempton 1986)

$$D_r \approx 100 \left(\frac{N_{60}}{60} \right)^{0.5} \qquad \text{(3-3b)}$$

where $D_r \geq 35$ percent. N_{60} should be multiplied by 0.92 for coarse sands and 1.08 for fine sands.

e. The undrained shear strength C_u in ksf may be estimated (Bowles 1988)

$$C_u \approx 0.12 N_{SPT} \qquad \text{(3-4)}$$

3. Cone penetration test (CPT). The *CPT* may be used to estimate both relative density of cohesionless soil and undrained strength of cohesive soils through empirical correlations. The *CPT* is especially suitable for sands and preferable to the *SPT*. The *CPT* may be performed using ASTM D 3441.

a. The relative density of several different sands can be estimated by (Jamiolkowski et al. 1988)

$$D_r = -74 + 66 \cdot \log_{10} \frac{q_c}{(\sigma'_{vo})^{0.5}} \qquad \text{(3-5)}$$

where the cone penetration resistance q_c and effective vertical overburden pressure σ'_{vo} are in units of ksf. The effective angle of internal friction ϕ' can be estimated from D_r using Table 3-1a. Table 3-1b provides a direct correlation of q_c with ϕ'.

b. The effective angle of internal friction decreases with increasing σ'_{vo} for a given q_c as approximately shown in Figure 3-1. Increasing confining pressure reduces ϕ' for a given q_c because the Mohr-Coulomb shear strengh envelope is nonlinear and has a smaller slope with increasing confining pressure.

c. The undrained strength C_u of cohesive soils can be estimated from (Schmertmann 1978)

$$C_u = \frac{q_c - \sigma_{vo}}{N_k} \qquad (3\text{-}6)$$

where C_u, q_c, and the total vertical overburden pressure σ_{vo} are in ksf units. The cone factor N_k should be determined using comparisons of C_u from laboratory undrained strength tests with the corresponding value of q_c obtained from the *CPT*. Equation 3-6 is useful to determine the distribution of undrained strength with depth when only a few laboratory undrained strength tests have been performed. N_k often varies from 14 to 20.

4. Dilatometer Test (DMT). The DMT can be used to estimate the overconsolidation ratio (*OCR*) distribution in the foundation soil. The *OCR* can be used in estimating the undrained strength. The *OCR* is estimated from the horizontal stress index K_D by (Baldi et al 1986; Jamiolkowski et al 1988)

$$OCR = (0.5K_D)^{1.56} \quad \text{if} \quad I_D \leq 1.2 \qquad (3\text{-}7a)$$

$$K_D = \frac{p_o - u_w}{\sigma'_{vo}} \qquad (3\text{-}7b)$$

$$I_D = \frac{p_1 - p_o}{p_1 - u_w} \qquad (3\text{-}7c)$$

where

p_o = internal pressure causing lift-off of the dilatometeter membrane, ksf

u_w = in situ hydrostatic pore pressure, ksf

p_1 = internal pressure required to expand the central point of the dilatometer membrane by $\approx$ 1.1 millimeters

K_D = horizontal stress index

I_D = material deposit index

The *OCR* typically varies from 1 to 3 for lightly overconsolidated soil and 6 to 8 for heavily overconsolidated soil.

5. Pressuremeter Test (PMT). The PMT can be used to estimate the undrained strength and the OCR. The PMT may be performed using ASTM D 4719.

a. The limit pressure p_L estimated from the PMT can be used to estimate the undrained strength by (Mair and Wood 1987)

$$C_u = \frac{p_L - \sigma_{ho}}{N_p} \qquad (3\text{-}8a)$$

$$N_p = 1 + 1n\frac{G_s}{C_u} \qquad (3\text{-}8b)$$

where

p_L = pressuremeter limit pressure, ksf

σ_{ho} = total horizontal in situ stress, ksf

G_s = shear modulus, ksf

p_L, σ_{ho}, and G_s are found from results of the PMT. Equation 3-8b requires an estimate of the shear strength to solve for N_p. N_p may be initially estimated as some integer value from 3 to 8 such as 6. The undrained strength is then determined from Equation 3-8a and the result substituted into Equation 3-8b. One or two iterations should be sufficient to evaluate C_u.

b. σ_{ho} can be used to estimate the *OCR* from $\sigma'_{ho}/\sigma'_{vo}$ if the pore water pressure and total vertical pressure distribution with depth are known or estimated.

6. Field Vane Shear Test (FVT). The FVT is commonly used to estimate the in situ undrained shear strength C_u of soft to firm cohesive soils. This test should be used with other tests when evaluating the soil shear strength. The test may be performed by hand or may be completed using sophisticated equipment. Details of the test are provided in ASTM D 2573.

a. The undrained shear strength C_u in ksf units is

$$C_u = \frac{T_v}{K_v} \qquad (3\text{-}9)$$

where

T_v = vane torque, kips · ft

K_v = constant depending on the dimensions and shape of the vane, ft^3

b. The constant K_v may be estimated for a rectangular vane causing a cylinder in a cohesive soil of uniform shear strength by

$$K_v = \frac{\pi}{1728} \cdot \frac{d_v^2 h_v}{2} \cdot \left[1 + \frac{d_v}{3h_v}\right] \qquad (3\text{-}10a)$$

where

d_v = measured diameter of the vane, in.

h_v = heasured height of the vane, in.

K_v for a tapered vane is

$$K_v = \frac{1}{1728} \cdot [\pi d_v^3 + 0.37(2d_v^3 - d_r^3)] \quad (3\text{-}10b)$$

where d_r is the rod diameter, in.

c. Anisotropy can significantly influence the torque measured by the vane.

D. WATER TABLE. Depth to the water table and pore water pressure distributions should be known to determine the influence of soil weight and surcharge on the bearing capacity as discussed in 1-4d, Chapter 1.

1. Evaluation of Groundwater Table (GWT). The GWT may be estimated in sands, silty sands, and sandy silts by measuring the depth to the water level in an augered hole at the time of boring and 24 hours thereafter. A 3/8 or 1/2 inch diameter plastic tube may be inserted in the hole for long-term measurements. Accurate measurements of the water table and pore water pressure distribution may be determined from piezometers placed at different depths. Placement depth should be within twice the proposed width of the foundation.

2. Fluctuations in GWT. Large seasonal fluctuations in GWT can adversely influence bearing capacity. Rising water tables reduce the effective stress in cohesionless soil and reduce the ultimate bearing capacity calculated using Equation 1-1.

3-3. Soil Exploration

Soil classification and index tests such as Atterberg Limit, gradations, and water content should be performed on disturbed soil and results plotted as a function of depth to characterize the types of soil in the profile. The distribution of shear strength with depth and the lateral variation of shear strength across the construction site should be determined from laboratory strength tests on undisturbed boring samples. Soil classifications and strengths may be checked and correlated with results of in situ tests. Refer to EM 1110-2-1907 and EM 1110-1-1804 for further information.

A. LATERAL DISTRIBUTION OF FIELD TESTS. Soil sampling, test pits, and in situ tests should be performed at different locations on the proposed site that may be most suitable for construction of the structure.

1. Accessibility. Accessibility of equipment to the construction site and obstacles in the construction area should be considered. It is not unusual to shift the location of the proposed structure on the construction site during soil exploration and design to accommodate features revealed by soil exploration and to achieve the functional requirements of the structure.

2. Location of Borings. Optimum locations for soil exploration may be near the center, edges, and corners of the proposed structure. A sufficient number of borings should be performed within the areas of proposed construction for laboratory tests to define shear strength parameters C_u and ϕ of each soil layer and any significant lateral variation in soil strength parameters for bearing capacity analysis and consolidation and compressibility characteristics for settlement analysis. These boring holes may also be used to measure water table depths and pore pressures for determination of effective stresses required in bearing capacity analysis.

a. Preliminary exploration should require two or three borings within each of several potential building locations. Air photos and geological conditions assist in determining location and spacings of borings along the alignment of proposed levees. Initial spacings usually vary from 200 to 1000 ft along the alignment of levees.

b. Detailed exploration depends on the results of the preliminary exploration. Eight to ten test borings within the proposed building area for typical structures are often required. Large and complex facilities may require more borings to properly define subsurface soil parameters. Refer to TM 5-818-1 for further information on soil exploration for buildings and EM 1110-2-1913 for levees.

B. DEPTH OF SOIL EXPLORATION. The depth of exploration depends on the size and type of the proposed structure and should be sufficient to assure that the soil supporting the foundation has adequate bearing capacity. Borings should penetrate all deposits which are unsuitable for foundation purposes such as unconsolidated fill, peat, loose sands, and soft or compressible clays.

1. Ten Percent Rule. The depth of soil exploration for at least one test boring should be at the depth where the increase in vertical stress caused by the structure is equal to 10 percent of the initial effective vertical overburden stress beneath the foundation

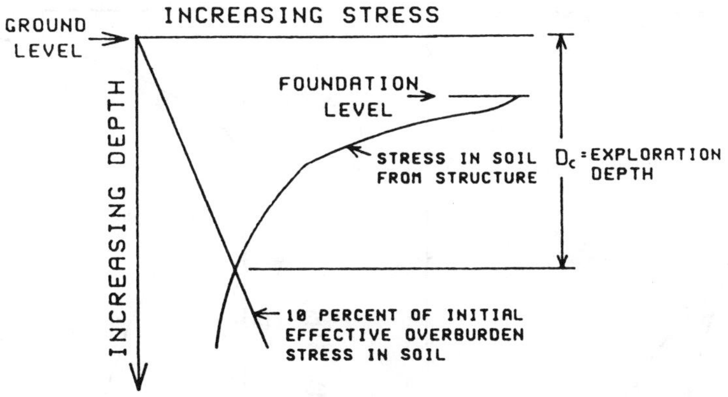

Figure 3-2. Estimation of the critical depth of soil exploration

(Figure 3-2). Critical depth for bearing capacity analysis D_c should be at least twice the minimum width of shallow square foundations or at least 4 times the minimum width of infinitely long footings or embankments. The depth of additional borings may be less if soil exploration in the immediate vicinity or the general stratigraphy of the area indicate that the proposed bearing strata have adequate thickness or are underlain by stronger formations.

2. Depth to Primary Formation. Depth of exploration need not exceed the depth of the primary formation where rock or soil of exceptional bearing capacity is located.

a. If the foundation is to be in soil or rock of exceptional bearing capacity, then at least one boring (or rock core) should be extended 10 or 20 ft into the stratum of exceptional bearing capacity to assure that bedrock and not boulders have been encountered.

b. For a building foundation carried to rock 3 to 5 rock corings are usually required to determine whether piles or drilled shafts should be used. The percent recovery and rock quality designation (RQD) value should be determined for each rock core. Drilled shafts are often preferred in stiff bearing soil and rock of good quality.

3. Selection of Foundation Depth. The type of foundation, whether shallow or deep, and the depth of undercutting for an embankment depends on the depths to acceptable bearing strata as well as on the type of structure to be supported.

a. Dense sands and gravels and firm to stiff clays with low potential for volume change provide the best bearing strata for foundations.

b. Standard penetration resistance values from the *SPT* and cone resistance from the *CPT* should be determined at a number of different lateral locations within the construction site. These tests should be performed to depths of about twice the minimum width of the proposed foundation.

c. Minimum depth requirements should be determined by such factors as depth of frost action, potential scour and erosion, settlement limitations, and bearing capacity.

C. SELECTION OF SHEAR STRENGTH PARAMETERS. Test data such as undrained shear strength C_u for cohesive soils and the effective angle of internal friction ϕ' for cohesionless sands and gravels should be plotted as a function of depth to determine the distribution of shear strength in the soil. Measurements or estimates of undrained shear strength of cohesive soils C_u are usually characteristic of the worst temporal case in which pore pressures build up in impervious foundation soil immediately following placement of structural loads. Soil consolidates with time under the applied foundation loads causing C_u to increase. Bearing capacity therefore increases with time.

1. Evaluation from Laboratory Tests. Undrained triaxial tests should be performed on specimens from undisturbed samples whenever possible to estimate strength parameters. The confining stresses of cohesive soils should be similar to that which will occur near potential failure planes in situ.

a. Effective stress parameters c', ϕ' may be evaluated from consolidatedundrained triaxial strength tests with pore pressure measurements (R) performed on undisturbed specimens according to EM 1110-2-1906. These specimens must be saturated.

b. The undrained shear strength C_u of cohesive foundation soils may be estimated from results of

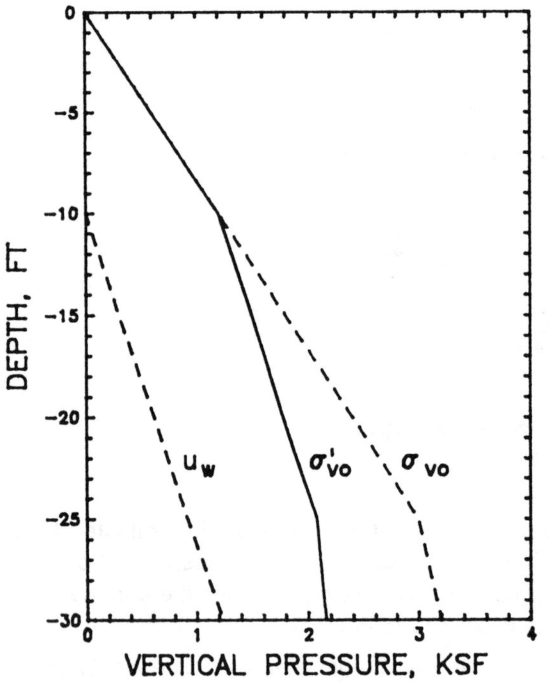

a. VERTICAL PRESSURE DISTRIBUTION

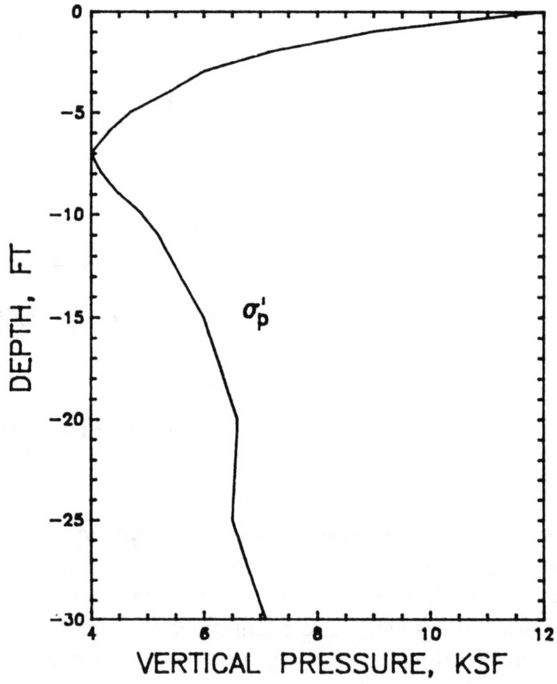

b. MAXIMUM PAST PRESSURE

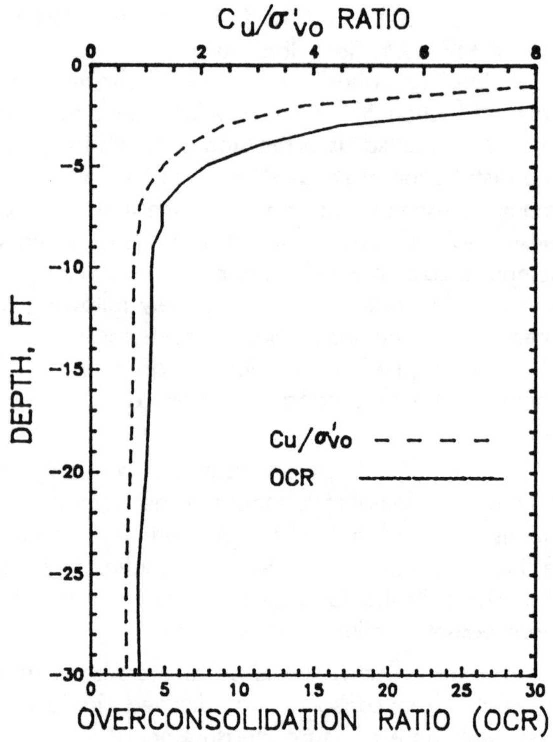

c. OVERCONSOLIDATION AND UNDRAINED COHESION RATIOS

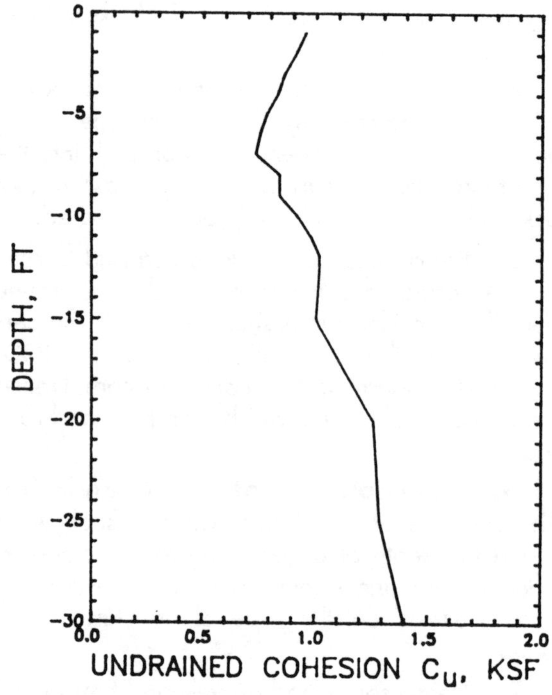

d. UNDRAINED COHESION DISTRIBUTION

Figure 3-3. Example estimation of undrained strength from maximum past pressure data

Table 3-3. Estimating Shear Strength of Soil From Maximum Past Pressure (Refer to Figure 3-3)

Step	Description
1.	Estimate the distribution of total vertical soil overburden pressure σ_{vo} with depth and make a plot as illustrated in Figure 3-3a.
2.	Estimate depth to groundwater table and plot the distribution of pore water pressure γ_w with depth, Figure 3-3a.
3.	Subtract pore water pressure distribution from the σ_{vo} distribution to determine the effective vertical soil pressure distribution σ'_{vo} and plot with depth, Figure 3-3a.
4.	Determine the maximum past pressure σ'_p from results of laboratory consolidation tests, in situ pressuremeter or other tests and plot with depth, Figure 3-3b.
5.	Calculate the overconsolidation ratio (*OCR*), σ'_p/σ'_{vo}, and plot with depth, Figure 3-3c.
6.	Estimate C_u/σ'_{vo} from $$\frac{C_u}{\sigma'_{vo}} = 0.25(OCR)^{0.8} \quad (3\text{-}11)$$ where C_u = undrained shear strength and plot with depth, Figure 3-3c.
7.	Calculate C_u by multiplying the ratio C_u/σ'_{vo} by σ'_{vo} and plot with depth, Figure 3-3d.
8.	An alternative approximation is $C_u \approx 0.2\sigma'_p$. For normally consolidated soils, $C_u/\sigma'_p = 0.11 + 0.0037 \cdot PI$, where *PI* is the plasticity index, percent (Terzaghi and Peck 1967).

unconsolidated-undrained (Q) triaxial tests according to EM 1110-2-1906 or standard test method ASTM D 2850. These tests should be performed on undrained undisturbed cohesive soil specimens at isotropic confining pressure similar to the total overburden pressure of the soil. Specimens should be taken from the center of undisturbed samples.

2. Estimates from Correlations. Strength parameters may be estimated by correlations with other data such as relative density, *OCR*, or the maximum past pressure.

a. The effective friction angle ϕ' of cohesionless soil may be estimated from in situ tests as described in Section 3-2c.

b. The distribution of undrained shear strength of cohesive soils may be roughly estimated from maximum past pressure soil data using the procedure outlined in Table 3-3. Pressure contributed by the foundation and structure are not included in this table, which increases conservatism of the shear strengths and avoids unnecessary complication of this approximate analysis. σ_{vo} refers to the total vertical pressure in the soil excluding pressure from any structural loads. σ'_{vo} is the effective vertical pressure found by subtracting the pore water pressure.

CHAPTER 4

SHALLOW FOUNDATIONS

4-1. Basic Considerations

Shallow foundations may consist of spread footings supporting isolated columns, combined footings for supporting loads from several columns, strip footings for supporting walls, and mats for supporting the entire structure.

A. SIGNIFICANCE AND USE. These foundations may be used where there is a suitable bearing stratum near the ground surface and settlement from compression or consolidation of underlying soil is acceptable. Potential heave of expansive foundation soils should also be acceptable. Deep foundations should be considered if a suitable shallow bearing stratum is not present or if the shallow bearing stratum is underlain by weak, compressible soil.

B. SETTLEMENT LIMITATIONS. Settlement limitation requirements in most cases control the pressure which can be applied to the soil by the footing. Acceptable limits for total downward settlement or heave are often 1 to 2 inches or less. Refer to EM 1110-1-1904 for evaluation of settlement or heave.

1. Total Settlement. Total settlement should be limited to avoid damage with connections in structures to outside utilities, to maintain adequate drainage and serviceability, and to maintain adequate freeboard of embankments. A typical allowable settlement for structures is 1 inch.

2. Differential Settlement. Differential settlement nearly always occurs with total settlement and must be limited to avoid cracking and other damage in structures. A typical allowable differential/span length ratio Δ/L for steel and concrete frame structures is 1/500 where Δ is the differential movement within span length L.

C. BEARING CAPACITY. The ultimate bearing capacity should be evaluated using results from a detailed in situ and laboratory study with suitable theoretical analyses given in 4-2. Design and allowable bearing capacities are subsequently determined according to Table 1-1.

4-2. Solution of Bearing Capacity

Shallow foundations such as footings or mats may undergo either a general or local shear failure. Local shear occurs in loose sands which undergo large strains without complete failure. Local shear may also occur for foundations in sensitive soils with high ratios of peak to residual strength. The failure pattern for general shear is modeled by Figure 1-3. Solutions of the general equation are provided using the Terzaghi, Meyerhof, Hansen and Vesic models. Each of these models have different capabilities for considering foundation geometry and soil conditions. Two or more models should be used for each design case when practical to increase confidence in the bearing capacity analyses.

A. GENERAL EQUATION. The ultimate bearing capacity of the foundation shown in Figure 1-6 can be determined using the general bearing capacity Equation 1-1

$$q_u = cN_c\zeta_c + \frac{1}{2}B'\gamma'_H N_\gamma \zeta_\gamma + \sigma'_D N_q \zeta_q \qquad (4\text{-}1)$$

where

q_u = ultimate bearing capacity, ksf

c = unit soil cohesion, ksf

B' = minimum effective width of foundation $B - 2e_B$, ft

e_B = eccentricity parallel with foundation width B, M_B/Q, ft

M_B = bending moment parallel with width B, kips-ft

Q = vertical load applied on foundation, kips

γ'_H = effective unit weight beneath foundation base within the failure zone, kips/ft^3

σ'_D = effective soil or surcharge pressure at the foundation depth D, $\gamma'_D \cdot D$, ksf

γ'_D = effective unit weight of soil from ground surface to foundation depth, kips/ft^3

D = foundation depth, ft

Table 4-1. Terzaghi dimensionless bearing capacity factors (after Bowles 1988)

ϕ'	N_q	N_c	N_γ	ϕ'	N_q	N_c	N_γ
28	17.81	31.61	15.7	0	1.00	5.70	0.0
30	22.46	37.16	19.7	2	1.22	6.30	0.2
32	28.52	44.04	27.9	4	1.49	6.97	0.4
34	36.50	52.64	36.0	6	1.81	7.73	0.6
35	41.44	57.75	42.4	8	2.21	8.60	0.9
36	47.16	63.53	52.0	10	2.69	9.60	1.2
38	61.55	77.50	80.0	12	3.29	10.76	1.7
40	81.27	95.66	100.4	14	4.02	12.11	2.3
42	108.75	119.67	180.0	16	4.92	13.68	3.0
44	147.74	151.95	257.0	18	6.04	15.52	3.9
45	173.29	172.29	297.5	20	7.44	17.69	4.9
46	204.19	196.22	420.0	22	9.19	20.27	5.8
48	287.85	258.29	780.1	24	11.40	23.36	7.8
50	415.15	347.51	1153.2	26	14.21	27.09	11.7

N_c, N_γ, N_q = dimensionless bearing capacity factors of cohesion c, soil weight in the failure wedge, and surcharge q terms

$\zeta_c, \zeta_\gamma, \zeta_q$ = dimensionless correction factors of cohesion c, soil weight in the failure wedge, and surcharge q accounting for foundation geometry and soil type

1. Net Bearing Capacity. The net ultimate bearing capacity q_u' is the maximum pressure that may be applied to the base of the foundation without undergoing a shear failure that is in addition to the overburden pressure at depth D.

$$q_u' = q_u - \gamma_D \cdot D \tag{4-2}$$

2. Bearing Capacity Factors. The dimensionless bearing capacity factors N_c, N_q, and N_γ are functions of the effective friction angle ϕ' and depend on the model selected for solution of Equation 4-1.

3. Correction Factors. The dimensionless correction factors ζ consider a variety of options for modeling actual soil and foundation conditions and depend on the model selected for solution of the ultimate bearing capacity. These options are foundation shape with eccentricity, inclined loading, foundation depth, foundation base on a slope, and a tilted foundation base.

B. TERZAGHI MODEL. An early approximate solution to bearing capacity was defined as general shear failure (Terzaghi 1943). The Terzaghi model is applicable to level strip footings placed on or near a level ground surface where foundation depth D is less than the minimum width B. Assumptions include use of a surface footing on soil at plastic equilibrium and a failure surface similar to Figure 1-3a. Shear resistance of soil above the base of an embedded foundation is not included in the solution.

1. Bearing Capacity Factors. The Terzaghi bearing capacity factors N_c and N_q for general shear are shown in Table 4-1 and may be calculated by

$$N_c = (N_q - 1)\cot \phi' \tag{4-3a}$$

$$N_q = \frac{e^{(270-\phi'/180)\pi \tan \phi'}}{2\cos^2(45 + \phi'/2)} \tag{4-3b}$$

Factor N_γ depends largely on the assumption of the angle ψ in Figure 1-3a. N_γ varies from minimum values using Hansen's solution to maximum values using the original Terzaghi solution. N_γ, shown in Table 4-1, was backfigured from the original Terzaghi values assuming $\psi = \phi'$ (Bowles 1988).

2. Correction Factors. The Terzaghi correction factors ζ_c and ζ_γ consider foundation shape only and are given in Table 4-2. ζ_q = 1.0 (Bowles 1988).

C. MEYERHOF MODEL. This solution considers correction factors for eccentricity, load inclination, and foundation depth. The influence of the shear strength of soil above the base of the foundation is considered in this solution. Therefore, beneficial effects of the foundation depth can be included in the analysis. Assumptions include use of a shape factor ζ_q

Table 4-2. Terzaghi correction factors ζ_c and ζ_γ

Factor	Strip	Square	Circular
ζ_c	1.0	1.3	1.3
ζ_γ	1.0	0.8	0.6

for surcharge, soil at plastic equilibrium, and a log spiral failure surface that includes shear above the base of the foundation. The angle $\psi = 45 + \phi/2$ was used for determination of N_γ. Table 4-3 illustrates the Meyerhof dimensionless bearing capacity and correction factors required for solution of Equation 4-1 (Meyerhof 1963).

1. Bearing Capacity Factors. Table 4-4 provides the bearing capacity factors in 2-degree intervals.

2. Correction Factors. Correction factors are given by

Cohesion: $\zeta_c = \zeta_{cs} \cdot \zeta_{ci} \cdot \zeta_{cd}$
Wedge: $\zeta_\gamma = \zeta_{\gamma s} \cdot \zeta_{\gamma i} \cdot \zeta_{\gamma d}$
Surcharge: $\gamma_q = \zeta_{qs} \cdot \zeta_{qi} \cdot \zeta_{qd}$

where subscript *s* indicates shape with eccentricity, subscript *i* indicates inclined loading, and *d* indicates foundation depth.

3. Eccentricity. The influence of bending moments on bearing capacity can be estimated by converting bending moments to an equivalent eccentricity *e*. Footing dimensions are then reduced to consider the adverse effect of eccentricity.

a. Effective footing dimensions may be given by

$$B' = B - 2e_B \quad (4\text{-}4a)$$

$$W' = W - 2e_W \quad (4\text{-}4b)$$

$$e_B = \frac{M_B}{Q} \quad (4\text{-}4c)$$

$$e_W = \frac{M_W}{Q} \quad (4\text{-}4d)$$

where

M_B = bending moment parallel with foundation width *B*, kips-ft

M_W = bending moment parallel with foundation length *W*, kips-ft

Orientation of axes, eccentricities,and bending moments are shown in Table 4-3.

b. The ultimate load applied to footings to cause a bearing failure is

$$Q_u = q_u \cdot A_e \quad (4\text{-}5)$$

where

q_u = ultimate bearing capacity of Equation 4-1 considering eccentricity in the foundation shape correction factor, Table 4-3, ksf

A_e = effective area of the foundation $B'W'$, ft^2

c. The bearing capacity of eccentric loaded foundations may also be estimated by (Meyerhof 1953)

$$q_{ue} = q_u \cdot R_e \quad (4\text{-}6)$$

where R_e is defined for cohesive soil by

$$R_e = 1 - 2 \cdot \frac{e}{B} \quad (4\text{-}7a)$$

and for cohesionless soil ($0 < e/B < 0.3$) by

$$R_e = 1 - \sqrt{\frac{e}{B}} \quad (4\text{-}7b)$$

where

q_u = ultimate capacity of a centrally-loaded foundation found from Equation 4-1 ignoring bending moments, ksf

e = eccentricity from Equations 4-4c and 4-4d, ft

D. HANSEN MODEL. The Hansen model considers tilted bases and slopes in addition to foundation shape and eccentricity, load inclination, and foundation depth. Assumptions are based on an extension of Meyerhof's work to include tilting of the base and construction on a slope. Any *D/B* ratio may be used permitting bearing capacity analysis of both shallow and deep foundations. Bearing capacity factors N_c and N_q are the same as Meyerhof's factors. N_γ is calculated assuming $\psi = 45 + \phi/2$. These values of N_γ are least of the methods. Correction factors ζ_c, ζ_γ, and ζ_q in Equation 4-1 are

Cohesion: $\zeta_c = \zeta_{cs} \cdot \zeta_{ci} \cdot \zeta_{cd} \cdot \zeta_{c\beta} \cdot \zeta_{c\delta}$ (4-8a)

Table 4-3. Meyerhof dimensionless bearing capacity and correction factors (data from Meyerhof 1953; Meyerhof 1963)

FACTOR			COHESION (c)	WEDGE (γ)	SURCHARGE (q)	DIAGRAM
BEARING CAPACITY N			N_c	N_γ	N_q	
		$\phi = 0$	5.14	0.00	1.00	
		$\phi > 0$	$(N_q - 1)\cot\phi$	$(N_q - 1)\tan(1.4\phi)$	$N_\phi e^{\pi \tan\phi}$	
CORRECTION ζ	FOUNDATION SHAPE WITH ECCENTRICITY s		ζ_{cs}	$\zeta_{\gamma s}$	ζ_{qs}	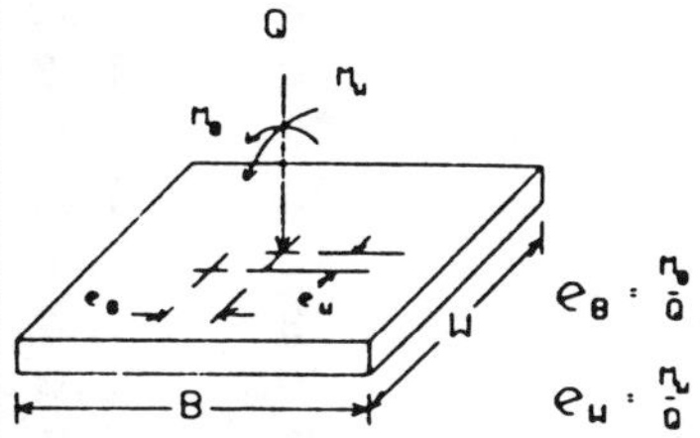
		$\phi = 0$	$1 + 0.2N_\phi \frac{B'}{W'}$	1.0	1.0	
		$\phi > 10$	"	$1 + 0.1N_\phi \frac{B'}{W'}$	$1 + 0.1N_\phi \frac{B'}{W'}$	
		$0 < \phi \le 10$	"	Linear Interpolation Between $\phi = 0$ and $\phi = 10$ Degrees		
	INCLINED LOADING i		ζ_{ci}	$\zeta_{\gamma i}$	ζ_{qi}	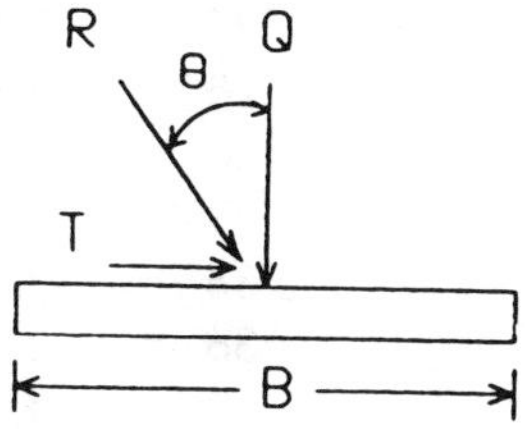
		$\phi = 0$	$\left[1 - \frac{\theta}{90}\right]$	1.0	$\left[1 - \frac{\theta}{90}\right]$	
		$\phi > 0$	$\left[1 - \frac{\theta}{90}\right]^2$	$\theta \le \phi$ $\left[1 - \frac{\theta}{\phi}\right]^2$; $\theta > \phi$ 0.0	$\left[1 - \frac{\theta}{90}\right]^2$	
	FOUNDATION DEPTH d		ζ_{cd}	$\zeta_{\gamma d}$	ζ_{qd}	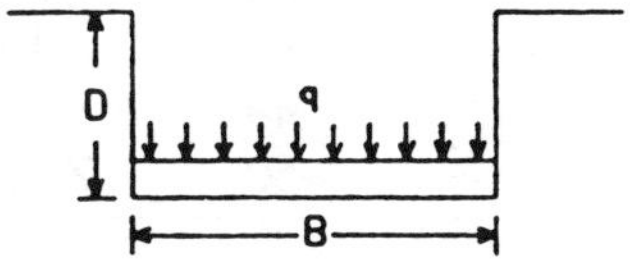
		$\phi = 0$	$1 + 0.2(N_\phi)^{1/2} \cdot \frac{D}{B}$	1.0	1.0	
		$\phi > 0$	"	$1 + 0.1(N_\phi)^{1/2} \cdot \frac{D}{B}$	$1 + 0.1(N_\phi)^{1/2} \cdot \frac{D}{B}$	
		$0 < \phi \le 10$	"	Linear Interpolation Between $\phi = 0$ and $\phi = 10$ Degrees		

Note: Eccentricity and inclined loading correction factors may not be used simultaneously; factors not used are unity

Nomenclature:
- ϕ = angle of internal friction, degrees
- N_ϕ = $\tan^2(45 + \phi/2)$
- B' = effective width of foundation, $B - 2e_B$, ft
- W' = effective lateral length of foundation, $W - 2e_W$, ft
- B = foundation width, ft
- W = foundation lateral length, ft
- D = foundation depth, ft
- Q = vertical load on foundation, qBW, kips
- q = bearing pressure on foundations, ksf
- T = horizontal load on foundation, right +, kips
- R = resultant load on foundation, $(Q^2 + T^2)^{1/2}$
- θ = angle of resultant load with vertical axis, $\cos^{-1}(Q/R)$, degrees
- e_B = eccentricity parallel with B, M_B/Q
- e_W = eccentricity parallel with W, M_W/Q
- M_B = bending moment parallel with B, kips-ft
- M_W = bending moment parallel with W, kips-ft

Wedge: $\zeta_\gamma = \zeta_{\gamma s} \cdot \zeta_{\gamma i} \cdot \zeta_{\gamma d} \cdot \zeta_{\gamma \beta} \cdot \zeta_{\gamma \delta}$ (4-8b)

Surcharge: $\zeta_q = \zeta_{qs} \cdot \zeta_{qi} \cdot \zeta_{qd} \cdot \zeta_{q\beta} \cdot \zeta_{q\delta}$ (4-8c)

where subscripts *s*, *i*, *d*, β, and δ indicate shape with eccentricity, inclined loading, foundation depth, ground slope, and base tilt, respectively. Table 4-5 illustrates the Hansen dimensionless bearing capacity and correction factors for solution of Equation 4-1.

Table 4-4. Meyerhof, Hansen, and Vesic dimensionless bearing capacity Factors

				N_γ		
ϕ	N_ϕ	N_c	N_q	Meyerhof	Hansen	Vesic
0	1.00	5.14	1.00	0.00	0.00	0.00
2	1.07	5.63	1.20	0.01	0.01	0.15
4	1.15	6.18	1.43	0.04	0.05	0.34
6	1.23	6.81	1.72	0.11	0.11	0.57
8	1.32	7.53	2.06	0.21	0.22	0.86
10	1.42	8.34	2.47	0.37	0.39	1.22
12	1.52	9.28	2.97	0.60	0.63	1.69
14	1.64	10.37	3.59	0.92	0.97	2.29
16	1.76	11.63	4.34	1.37	1.43	3.06
18	1.89	13.10	5.26	2.00	2.08	4.07
20	2.04	14.83	6.40	2.87	2.95	5.39
22	2.20	16.88	7.82	4.07	4.13	7.13
24	2.37	19.32	9.60	5.72	5.75	9.44
26	2.56	22.25	11.85	8.00	7.94	12.54
28	2.77	25.80	14.72	11.19	10.94	16.72
30	3.00	30.14	18.40	15.67	15.07	22.40
32	3.25	35.49	23.18	22.02	20.79	30.21
34	3.54	42.16	29.44	31.15	28.77	41.06
36	3.85	50.59	37.75	44.43	40.05	56.31
38	4.20	61.35	48.93	64.07	56.17	78.02
40	4.60	75.31	64.19	93.69	79.54	109.41
42	5.04	93.71	85.37	139.32	113.95	155.54
44	5.55	118.37	115.31	211.41	165.58	224.63
46	6.13	152.10	158.50	328.73	244.64	330.33
48	6.79	199.26	222.30	526.44	368.88	495.99
50	7.55	266.88	319.05	873.84	568.56	762.85

1. Restrictions. Foundation shape with eccentricity ζ_{cs}, $\zeta_{\gamma s}$, and ζ_{qs} and inclined loading ζ_{ci}, $\zeta_{\gamma i}$, and ζ_{qi} correction factors may not be used simultaneously. Correction factors not used are unity.

2. Eccentricity. Influence of bending moments is evaluated as in the Meyerhof model.

3. Inclined loads. The *B* component in Equation 4-1 should be width *B* if horizontal load *T* is parallel with *B* or should be *W* if *T* is parallel with length *W*.

E. VESIC MODEL. Table 4-6 illustrates the Vesic dimensionless bearing capacity and correction factors for solution of Equation 4-1.

1. Bearing Capacity Factors. N_c and N_q are identical with Meyerhof's and Hansen's factors. N_γ was taken from an analysis by Caquot and Kerisel (1953) using $\psi = 45 + \phi/2$.

2. Local Shear. A conservative estimate of N_q may be given by

$$N_q = (1 + \tan\phi') \cdot e^{\tan\phi'} \cdot \tan^2\left[45 + \frac{\phi'}{2}\right] \quad (4\text{-}9)$$

Equation 4-9 assumes a local shear failure and leads to a lower bound estimate of q_u. N_q from Equation 4-9 may also be used to calculate N_c and N_γ by the equations given in Table 4-6.

F. COMPUTER SOLUTIONS. by computer programs provide effective methods of estimating ultimate and allowable bearing capacities.

1. Program CBEAR. Program CBEAR (Mosher and Pace 1982) can be used to calculate the bearing capacity of shallow strip, rectangular, square, or circular footings on one or two soil layers. This program uses the Meyerhof and Vesic bearing capacity factors and correction factors.

Table 4-5. Hansen dimensionless bearing capacity and correction factors (data from Hansen 1970)

FACTOR			COHESION (c)	WEDGE (γ)	SURCHARGE (q)	DIAGRAM
BEARING CAPACITY N			N_c	N_γ	N_q	
		$\phi = 0$	5.14	0.00	1.00	
		$\phi > 0$	$(N_q-1)\cot\phi$	$1.5(N_q-1)\tan\phi$	$N_\phi e^{\pi\tan\phi}$	
CORRECTION ζ	FOUNDATION SHAPE WITH ECCENTRICITY s		ζ_{cs}	$\zeta_{\gamma s}$	ζ_{qs}	
			Strip: 1.0			
		$\phi = 0$	$0.2\cdot\frac{B'}{W'}$	1.0	1.0	
		$\phi > 0$	$1+\frac{N_q B'}{N_c W'}$	$1-0.4\cdot\frac{B'}{W'}$	$1+\frac{B'}{W'}\tan\phi$	
	INCLINED LOADING i		ζ_{ci}	$\zeta_{\gamma i}$	ζ_{qi}	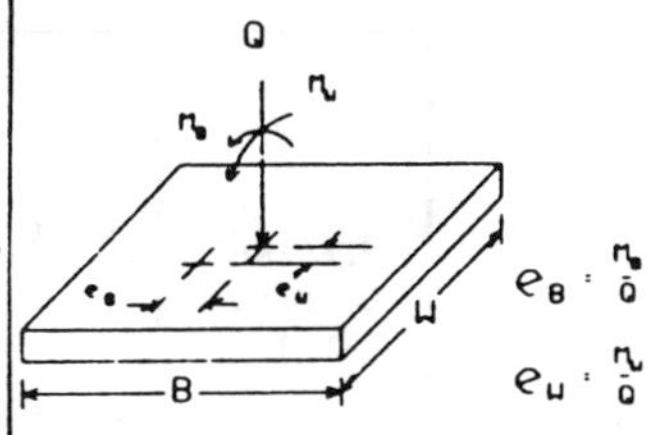
		$\phi = 0$	$\frac{1-\left[1-\frac{T}{A_e c_a}\right]^{\frac{1}{2}}}{2}$	$\delta=0$ $\left[1-\frac{0.7T}{Q+A_e c_a\cot\phi}\right]^5$	$\left[1-\frac{0.5T}{Q+A_e c_a\cot\phi}\right]^5$	
		$\phi > 0$	$\zeta_{qi}-\frac{1-\zeta_{qi}}{N_q-1}$	$\delta>0$ $\left[1-\frac{(0.7-\delta/450)T}{Q+A_e c_a\cot\phi}\right]^5$		
	FOUNDATION DEPTH d		ζ_{cd}	$\zeta_{\gamma d}$	ζ_{qd}	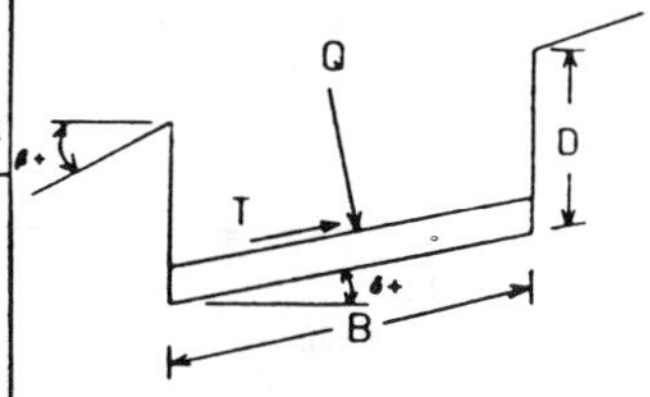
		$\phi = 0$	0.4k	1.0	1.0	
		$\phi > 0$	1 + 0.4k	1.0	$1+2\tan\phi(1-\sin\phi)^2 k$	
	BASE ON SLOPE β		$\zeta_{c\beta}$	$\zeta_{\gamma\beta}$	$\zeta_{q\beta}$	
		$\phi = 0$	$1-\frac{\beta}{147.3}$	$(1-0.5\tan\beta)^5$	$(1-0.5\tan\beta)^5$	$\beta+\delta \le 90°$
		$\phi > 0$	$\zeta_{q\beta}-\frac{1-\zeta_{q\beta}}{147.3}$			$\beta < \phi$
	TILTED BASE δ		$\zeta_{c\delta}$	$\zeta_{\gamma\delta}$	$\zeta_{q\delta}$	
		$\phi = 0$	$1-\frac{\delta}{147}$	$e^{-0.047\delta\tan\phi}$	$e^{-0.035\delta\tan\phi}$	
		$\phi > 0$	$\zeta_{q\delta}-\frac{1-\zeta_{q\delta}}{147.3}$			

Note: Eccentricity and inclined loading correction factors may not be used simultaneously; factors not used are unity

Nomenclature:

- ϕ = angle of internal friction, degrees
- N_ϕ = $\tan^2(45+\phi/2)$
- B' = effective width of foundation, $B - 2e_B$, ft
- W' = effective length of foundation, $W - 2e_W$, ft
- B = foundation width, ft
- W = foundation length, ft
- e_B = eccentricity parallel with B, M_B/Q
- e_W = eccentricity parallel with W, M_W/Q
- M_B = bending moment parallel with B, kips-ft
- M_W = bending moment parallel with W, kips-ft
- ϕ_a = friction angle between base and soil $\approx \phi$, degrees
- c_a = adhesion of soil to base $\le$ c, ksf
- c = soil cohesion or undrained shear strength C_u, ksf
- δ = base tilt from horizontal, upward +, degrees
- β = slope of ground from base, downward +, degrees
- k = D/B if $D/B \le 1$ OR $\tan^{-1}(D/B)$ if $D/B > 1$ (in radians)
- D = foundation depth, ft
- Q = vertical load on foundation, kips
- T = horizontal load $\le Q\tan\phi + A_e c_a$, right +, kips
- A_e = effective area of foundation $B'W'$, ft^2

2. Program UTEXAS2. UTEXAS2 is a slope stability program that can be used to calculate factors of safety for long wall footings and embankments consisting of multilayered soils (Edris 1987). Foundation loads are applied as surface pressures on flat surfaces or slopes. Circular or noncircular failure surfaces may be assumed. Noncircular failure surfaces may be straight lines and include wedges. Shear surfaces are directed to the left of applied surface loading on horizontal slopes or in the direction in which gravity

Table 4-6. Vesic dimensionless bearing capacity and correction factors (data from Vesic 1973; Vesic 1975)

FACTOR			COHESION (c)	WEDGE (γ)	SURCHARGE (q)	DIAGRAM
BEARING CAPACITY N			N_c	N_γ	N_q	
		$\phi = 0$	5.14	0.00 OR $-2\sin\beta$ if $\beta>0$	1.00	
		$\phi > 0$	$(N_q-1)\cot\phi$	$2(N_q+1)\tan\phi$	$N_\phi e^{\pi\tan\phi}$	
CORRECTION ζ	FOUNDATION SHAPE WITH ECCENTRICITY s		ζ_{cs}	$\zeta_{\gamma s}$	ζ_{qs}	
			Strip: 1.0			
		$\phi = 0$	$0.2 \cdot \frac{B'}{W'}$	1.0	1.0	
		$\phi > 0$	$1 + \frac{N_q B'}{N_c W'}$	$1 - 0.4 \cdot \frac{B'}{W'}$ (1.0 if strip)	$1 + \frac{B'}{W'}\tan\phi$ (1.0 if strip)	
	INCLINED LOADING i		ζ_{ci}	$\zeta_{\gamma i}$	ζ_{qi}	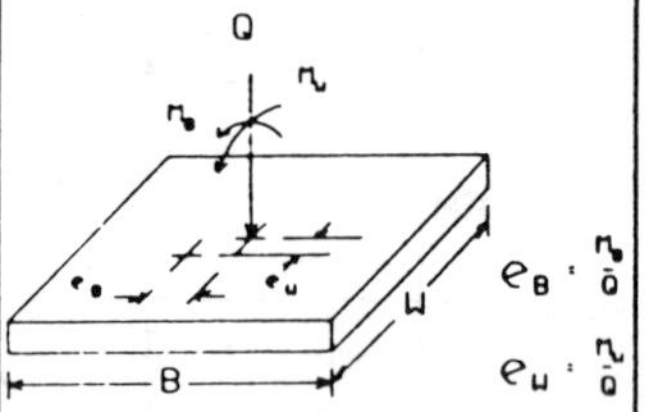
		$\phi = 0$	$\frac{1 - \frac{mT}{A_e c_a N_c}}{2}$	$\left[1 - \frac{T}{Q+A_e c_a \cot\phi}\right]^{m+1} > 0$	$\left[1 - \frac{T}{Q+A_e c_a \cot\phi}\right]^{m}$	
		$\phi > 0$	$\zeta_{qi} - \frac{1 - \zeta_{qi}}{N_q - 1}$			
	FOUNDATION DEPTH d		ζ_{cd}	$\zeta_{\gamma d}$	ζ_{qd}	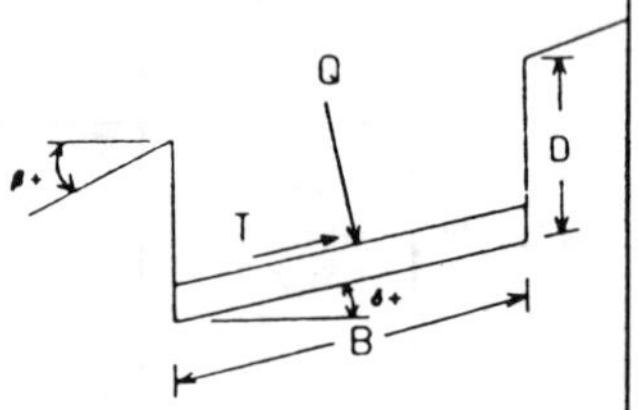
		$\phi = 0$	$1 + 0.4K$	1.0	1.0	
		$\phi > 0$	$1 + 0.4k$	1.0	$1 + 2\tan\phi(1-\sin\phi)^2 k$	
	BASE ON SLOPE β		$\zeta_{c\beta}$	$\zeta_{\gamma\beta}$	$\zeta_{q\beta}$	
		$\phi = 0$	$1 - \frac{\beta}{147.3}$	$(1 - \tan\beta)^2$	$(1 - \tan\beta)^2$	$\beta + \delta \le 90°$
		$\phi > 0$	$\zeta_{q\beta} - \frac{1 - \zeta_{q\beta}}{147.3}$			$\beta < \phi$
	TILTED BASE δ		$\zeta_{c\delta}$	$\zeta_{\gamma\delta}$	$\zeta_{q\delta}$	
		$\phi = 0$	$1 - \frac{\delta}{147}$	$(1 - 0.017\delta\tan\phi)^2$	$(1 - 0.017\delta\tan\phi)^2$	
		$\phi > 0$	$\zeta_{q\delta} - \frac{1 - \zeta_{q\delta}}{147.3}$			

Note: Eccentricity and inclined loading correction factors may not be used simultaneously; factors not used are unity.

Nomenclature:

- ϕ = angle of internal friction, degrees
- N_ϕ = $\tan^2(45 + \phi/2)$
- B' = effective width of foundation, $B - 2e_B$, ft
- W' = effective length of foundation, $W - 2e_W$, ft
- B = foundation width, ft
- W = foundation length, ft
- e_B = eccentricity parallel with B, M_B/Q
- e_W = eccentricity parallel with W, M_W/Q
- M_B = bending moment parallel with B, kips-ft
- M_W = bending moment parallel with W, kips-ft
- ϕ_a = friction angle between base and soil ≈ ϕ, degrees
- c_a = adhesion of soil to base ≤ c, ksf
- c = soil cohesion or undrained shear strength C_u, ksf
- δ = base tilt from horizontal, upward +, degrees
- β = slope of ground from base, downward +, degrees
- k = D/B if D/B ≤ 1 OR $\tan^{-1}(D/B)$ if D/B > 1 (in radians)
- D = foundation depth, ft
- Q = vertical load on foundation, kips
- T = horizontal load ≤ $Q\tan\phi + A_e c_a$, right +, kips
- A_e = effective area of foundation $B'W'$, ft^2
- m = $\frac{2 + R_{BW}}{1 + R_{BW}}$, $R_{BW} = B/W$ if T parallel to B; $R_{BW} = W/B$ if T parallel to W

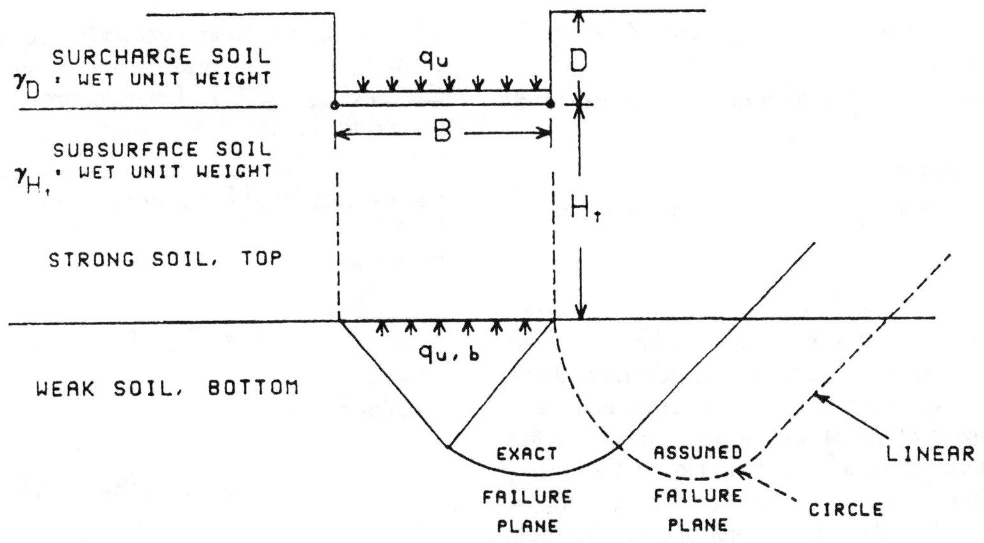

Figure 4-1. Schematic of a multilayer foundation-soil system

would produce sliding on nonhorizontal slopes (e.g., from high toward low elevation points). This program can also consider the beneficial effect of internal reinforcement in the slope. q_u calculated by UTEXAS2 may be different from that calculated by CBEAR partly because the *FS* is defined in UTEXAS2 as the available shear strength divided by the shear stress on the failure surface. The assumed failure surfaces in CBEAR are not the same as the minimum *FS* surface found in UTEXAS2 by trial and error. *FS* in Table 1-2 are typical for CBEAR. Program UTEXAS2 calculates factors of safety, but these *FS* have not been validated with field experience. UTEXAS2 is recommended as a supplement to the Terzaghi, Meyerhof, Hansen, and Vesic models until *FS* determined by UTEXAS2 have been validated.

G. MULTILAYER SOILS. Foundations are often supported by multilayer soils. Multiple soil layers influence the depth of the failure surface and the calculated bearing capacity. Solutions of bearing capacity for a footing in a strong layer that is overlying a weak clay layer, Figure 4-1, are given below. These solutions are valid for a punching shear failure. The use of more than two soil layers to model the subsurface soils is usually not necessary.

1. Depth of Analysis. The maximum depth of the soil profile analyzed need not be much greater than the depth to the failure surface, which is approximately 2B for uniform soil. A deeper depth may be required for settlement analyses.

a. If the soil immediately beneath the foundation is weaker than deeper soil, the critical failure surface may be at a depth less than 2*B*.

b. If the soil is weaker at depths greater than 2*B*, then the critical failure surface may extend to depths greater than 2*B*.

2. Dense Sand Over Soft Clay. The ultimate bearing capacity of a footing in a dense sand over soft clay can be calculated assuming a punching shear failure using a circular slip path (Hanna and Meyerhof 1980; Meyerhof 1974)

Wall Footing:

$$q_u = q_{u,b} + \frac{2\gamma_{sand}H_t^2}{B}\left(1 + \frac{2D}{H_t}\right)K_{ps}\tan\phi_{sand} - \gamma_{sand}H_t \leq q_{ut} \quad \text{(4-10a)}$$

Circular Footing:

$$q_u = q_{u,b} + \frac{2\gamma_{sand}H_t^2}{B}\left(1 + \frac{2D}{H_t}\right)S_sK_{ps}\tan\phi_{sand} - \gamma_{sand}H_t \leq q_{ut} \quad \text{(4-10b)}$$

where

$q_{u,b}$ = ultimate bearing capacity on a very thick bed of the bottom soft clay layer, ksf

γ_{sand} = wet unit weight of the upper dense sand, kips/ft^3

H_t = depth below footing base to soft clay, ft

D = depth of footing base below ground surface, ft

K_{ps} = punching shear coefficient, Figure 4-2a, 4-2b, and 4-2c

ϕ_{sand} = angle of internal friction of upper dense sand, degrees

S_s = shape factor

q_{ut} = ultimate bearing capacity of upper dense sand, ksf

The punching shear coefficient k_{ps} can be found from the charts in Figure 4-2 using the undrained shear strength of the lower soft clay and a punching shear parameter C_{ps}. C_{ps}, ratio of ζ/ϕ_{sand} where ζ is the average mobilized angle of shearing resistance on the assumed failure plane, is found from Figure 4-2d using ϕ_{sand} and the bearing capacity ratio R_{bc}. $R_{bc} = 0.5\gamma_{sand}BN_\gamma/(C_uN_c)$. B is the diameter of a circular footing or width of a wall footing. The shape factor S_s, which varies from 1.1 to 1.27, may be assumed unity for conservative design.

3. Stiff Over Soft Clay. Punching shear failure is assumed for stiff over soft clay.

a. $D = 0.0$. The ultimate bearing capacity can be calculated for a footing on the ground surface by (Brown and Meyerhof 1969)

Wall Footing:

$$q_u = C_{u,upper}N_{cw,0} \qquad (4\text{-}11a)$$

Circular Footing:

$$N_{cw,0} = 1.5\frac{H_t}{B_{dia}} + 5.14\frac{C_{u,lower}}{C_{u,upper}} \leq 5.14 \qquad (4\text{-}11b)$$

$$q_u = C_{u,upper}N_{cc,0} \qquad (4\text{-}11c)$$

$$N_{cc,0} = 3.0\frac{H_t}{B_{dia}} + 6.05\frac{C_{u,lower}}{C_{u,upper}} \leq 6.05 \qquad (4\text{-}11d)$$

where

$C_{u,upper}$ = undrained shear strength of the stiff upper clay, ksf

$C_{u,lower}$ = undrained shear strength of the soft lower clay, ksf

$N_{cw,0}$ = bearing capacity factor of the wall footing

$N_{cc,0}$ = bearing capacity factor of the circular footing

B_{dia} = diameter of circular footing, ft

A rectangular footing may be converted to a circular footing by $B_{dia} = 2(BW/\pi)^{1/2}$ where B = width and W = length of the footing. Factors $N_{cw,0}$ and $N_{cc,0}$ will overestimate bearing capacity by about 10 percent if $C_{u,lower}/C_{u,upper} \geq 0.7$. Refer to Brown and Meyerhof (1969) for charts of $N_{cw,0}$ and $N_{cc,0}$.

b. $D > 0.0$. The ultimate bearing capacity can be calculated for a footing placed at depth D by

Wall Footing:

$$q_u = C_{u,upper}N_{cw,D} + \gamma D \qquad (4\text{-}12a)$$

Circular Footing:

$$q_u = C_{u,upper}N_{cc,D} + \gamma D \qquad (4\text{-}12b)$$

where

$N_{cw,D}$ = bearing capacity factor of wall footing with $D > 0.0$

$N_{cc,D}$ = bearing capacity factor of rectangular footing with $D > 0.0$

$= N_{cw,D}[1 + 0.2(B/W)]$

γ = wet unit soil weight of upper soil, kips/ft^3

D = depth of footing, ft

$N_{cw,D}$ may be found using Table nd $N_{cw,0}$ from Equation 4-11b. Refer to Department of the Navy (1982) for charts that can be used to calculate bearing capacities in two layer soils.

4. Computer Analysis. The bearing capacity of multilayer soils may be estimated from computer solutions using program CBEAR (Mosher and Pace 1982). Program UTEXAS2 (Edris 1987) calculates *FS* for wall footings and embankments, which have not been validated with field experience. UTEXAS2 is recommended as a supplement to CBEAR until *FS* have been validated.

H. CORRECTION FOR LARGE FOOTINGS AND MATS. Bearing capacity, obtained using Equation 4-1 and the bearing capacity factors, gives capacities that are too large for widths $B > 6$ ft. This is apparently because the $0.5 \cdot B'\gamma'_H N_\gamma \zeta_\gamma$ term becomes too large (DeBeer 1965; Vesic 1969).

1. Settlement usually controls the design and loading of large dimensioned structures because the foundation soil is stressed by the applied loads to deep depths.

2. Bearing capacity may be corrected for large footings or mats by multiplying the surcharge

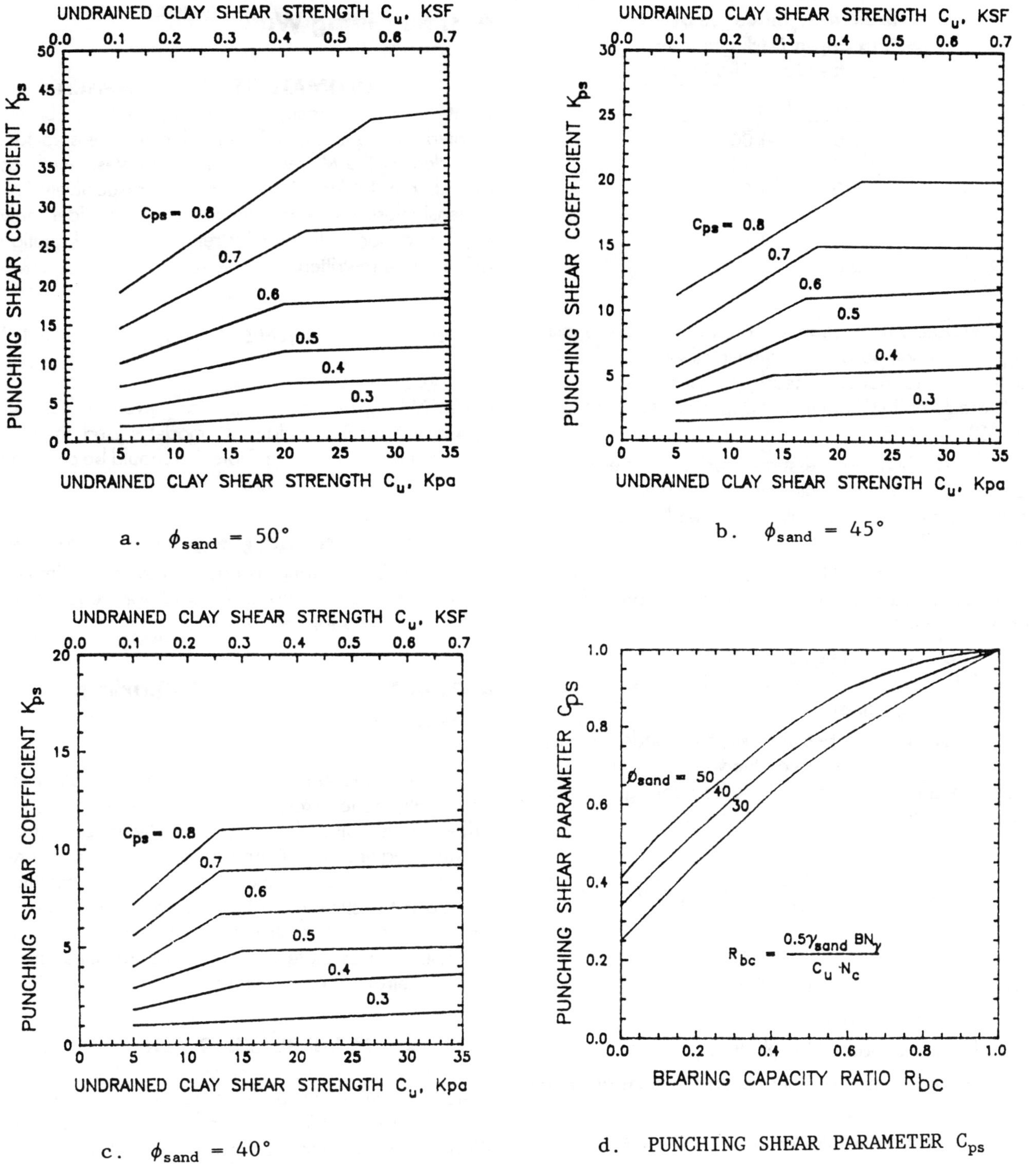

Figure 4-2. Charts for calculation of ultimate bearing capacity of dense sand over soft clay (Data from Hanna and Meyerhof 1980)

term $0.5 \cdot B'\gamma'_H N_\gamma \zeta_\gamma$ by a reduction factor (Bowles 1988)

$$r_\gamma = 1 - 0.25\log_{10}\frac{B}{6} \qquad (4\text{-}13)$$

where $B > 6$ ft.

I. PRESUMPTIVE BEARING CAPACITY.

Refer to Table or typical presumptive allowable bearing pressures q_{na}. Presumptive allowable pressures should

Table 4-7. Influence of footing depth *D* (department of the Navy 1982)

D/B	$N_{cw,D}/N_{cw,0}$
0.0	1.00
0.5	1.15
1.0	1.24
2.0	1.36
3.0	1.43
4.0	1.46

only be used with caution for spread footings supporting small or temporary structures and verified, if practical, by performance of nearby structures. Further details are given in Chapter 4 of Department of the Navy (1982).

1. Bearing pressures produced by eccentric loads that include dead plus normal live loads plus permanent lateral loads should not exceed q_{na} pressures of Table 4-8.

2. Transient live loads from wind and earthquakes may exceed the allowable bearing pressure by up to one-third.

3. For footings of width $B < 3$ ft in least lateral dimension the allowable bearing pressures is B times 1/3 of q_{na} given in Table 4-8.

4. For a bearing stratum underlain by weaker material, pressure on the weak stratum should be less than the nominal allowable bearing pressure given in Table 4-8

$$\frac{Q}{(B + 1.16H_t) \cdot (W + 1.16H_t)} \leq q_{na} \quad (4\text{-}14)$$

where

Q = vertical load on foundation, kips

B = foundation width, ft

W = foundation lateral length, ft

H_t = depth to weak stratum beneath bottom of foundation, ft

q_{na} = nominal allowable bearing pressure, ksf

5. Resistance to uplift force Q_{up} should be

$$\frac{W'_T}{Q_{up}} > 2 \quad (4\text{-}15)$$

where W'_T is the total effective weight of soil and foundation resisting uplift.

4-3. Retaining Walls

A. ULTIMATE BEARING CAPACITY. Ultimate bearing capacity of retaining walls may be estimated by Equation 4-1 with dimensionless factors provided by the Meyerhof, Hansen, or Vesic methods described in Tables 4-3, 4-5, and 4-6, respectively. The dimensionless correction factors need consider only depth and load inclination for retaining walls. Equation 4-1 may be rewritten

$$q_u = cN_c\zeta_{cd}\zeta_{ci} + \frac{1}{2}B'\gamma'_H N_\gamma \zeta_{\gamma d}\zeta_{\gamma i} + \sigma'_D N_q \zeta_{qd}\zeta_{qi} \quad (4\text{-}16)$$

where N_c, N_γ, N_q and ζ_c, ζ_γ, ζ_q are given in Tables 4-3, 4-4, 4-5, or 4-6. If Hansen's model is used, then the exponent for $\zeta_{\gamma i}$ and ζ_{qi} in Table 4-5 should be changed from 5 to 2 (Bowles 1988).

B. ALLOWABLE BEARING CAPACITY. The allowable bearing capacity may be estimated from Equations 1-2 using $FS = 2$ for cohesionless soils and $FS = 3$ for cohesive soils.

4-4. In Situ Modeling of Bearing Pressures

In situ load tests of the full size foundation are not usually done, except for load testing of piles and drilled shafts. Full scale testing is usually not performed because required loads are usually large and as a result these tests are expensive. The most common method is to estimate the bearing capacity of the soil from the results of relatively simple, less expensive in situ tests such as plate bearing, standard penetration, cone penetration, and vane shear tests.

A. PLATE BEARING TEST. Loading small plates 12 to 30 inches in diameter or width B_p are quite useful, particularly in sands, for estimating the bearing capacity of foundations. The soil strata within a depth 4B beneath the foundation must be similar to the strata beneath the plate. Details of this test are described in standard method ASTM D 1194. A large vehicle can be used to provide reaction for the applied pressures.

1. Constant Strength. The ultimate bearing capacity of the foundation in cohesive soil of constant shear strength may be estimated by

$$B < 4B_p\text{: } q_u = q_{u,p} \quad (4\text{-}17a)$$

Table 4-8. Presumptive Allowable Bearing Pressures for Spread Footings (Data from Department of the Navy 1982, Table 1, Chapter 4)

Bearing Material	In Place Consistency	Nominal Allowable Bearing Pressure q_{na}, ksf
Massive crystalline igneous and metamorphic rock: granite, diorite, basalt, gneiss, thoroughly cemented conglomerate (sound condition allows minor cracks)	Hard sound rock	160
Foliated metamorphic rock: slate, schist (sound condition allows minor cracks)	Medium hard sound rock	70
Sedimentary rock; hard cemented shales, siltstone, sandstone, limestone without cavities	Medium hard sound rock	40
Weathered or broken bed rock of any kind except highly argillaceousrock (shale); Rock Quality Designation less than 25	Soft rock	20
Compaction shale or other highly argillaceous rock in sound condition	Soft rock	20
Well-graded mixture of fine and coarse-grained soil: glacial till, hardpan, boulder clay (*GW-GC, GC, SC*)	Very compact	20
Gravel, gravel-sand mixtures, boulder-gravel mixtures (*SW, SP, SW, SP*)	Very compact	14
	Medium to compact	10
	Loose	6
Coarse to medium sand, sand with little gravel (*SW, SP*)	Very compact	8
	Medium to compact	6
	Loose	3
Fine to medium sand, silty or clayey medium to coarse sand (*SW, SM, SC*)	Very compact	6
	Medium to compact	5
	Loose	3
Homogeneous inorganic clay, sandy or silty clay (*CL, CH*)	Very stiff to hard	8
	Medium to stiff	4
	Soft	1
Inorganic silt, sandy or clayey silt, varved silt-clay-fine sand	Very stiff to hard	6
	Medium to stiff	3
	Soft	1

where

q_u = ultimate bearing capacity of the foundation, ksf

$q_{u,p}$ = ultimate bearing capacity of the plate, ksf

B = diameter or width of the foundation, ft

B_p = diameter or width of the plate, ft

2. Strength Increasing Linearly With Depth. The ultimate bearing capacity of the foundation in cohesionless or cohesive soil with strength increasing linearly with depth may be estimated by

$$B < 4B_p\text{: } q_u = q_{u,p}\,\frac{B}{B_p} \tag{4-17b}$$

3. Extrapolation of Settlement Test Results in Sands. The soil pressure q_1 may be estimated using a modified Terzaghi and Peck approximation (Peck and Bazarra 1969; Peck, Hanson, and Thornburn 1974)

$$q_1 = \frac{q}{1.5}\,\rho_i \tag{4-18}$$

where

q_1 = soil pressure per inch of settlement, ksf/in.

q = average pressure applied on plate, ksf

ρ_i = immediate settlement of plate, in.

The results of the plate load test should indicate that q/ρ_i is essentially constant. q_1 and plate diameter B_p can then be input into the Terzaghi and Peck chart for the appropriate D/B ratio, which is 1, 0.5 or 0.25 (see Figure 3-3, EM 1110-1-1904). The actual footing dimension B is subsequently input into the same chart to indicate the allowable foundation bearing pressure.

4. Extrapolation of Test Results. Load tests performed using several plate sizes may allow extrapolation of test results to foundations up to 6 times the plate diameter provided the soil is similar. Other in situ test results using standard penetration or cone penetration data are recommended for large foundation diameters and depths more than $4B_p$.

B. STANDARD PENETRATION TEST (SPT). The SPT may be used to directly obtain the allowable bearing capacity of soils for specific amounts of settlement based on past correlations.

1. Footings. Meyerhof's equations (Meyerhof 1956; Meyerhof 1974) are modified to increase bearing capacity by 50 percent (Bowles 1988)

$$B \leq 4 \text{ ft:} \qquad q_{a,1} = \frac{N_n}{F_1} K_d \tag{4-19a}$$

$$B > 4 \text{ ft:} \qquad q_{a,1} = \frac{N_n}{F_2}\left[\frac{B + F_3}{B}\right]^2 \tag{4-19b}$$

where

$q_{a,1}$ = allowable bearing capacity for 1 inch of settlement, ksf

$K_d = 1 + 0.33(D/B) \leq 1.33$

N_n = standard penetration resistance corrected to n percent energy

Equation 4-19b may be used for footings up to 15 ft wide.

a. F factors depend on the energy of the blows. n is approximately 55 percent for uncorrected penetration resistance and $F_1 = 2.5$, $F_2 = 4$, and $F_3 = 1$. F factors corrected to $n = 70$ percent energy are $F_1 = 2$, $F_2 = 3.2$ and $F_3 = 1$.

b. Figure 3-3 of EM 1110-1-1904 provides charts for estimating q_a for 1 inch of settlement from SPT data using modified Terzaghi and Peck approximations.

2. Mats. For mat foundations

$$q_{a,1} = \frac{N_n}{F_2} K_d \tag{4-20a}$$

where $q_{a,1}$ is the allowable bearing capacity for limiting settlement to 1 inch. The allowable bearing capacity for any settlement q_a may be linearly related to the allowable settlement for 1 inch obtained from Equations 4-19 assuming settlement varies in proportion to pressure

$$q_a = \rho \cdot q_{a,1} \tag{4-20b}$$

where

ρ = settlement, inches

$q_{a,1}$ = allowable bearing capacity for 1 inch settlement, ksf

C. CONE PENETRATION TEST (CPT). Bearing capacity has been correlated with cone tip resistance q_c for shallow foundations with $D/B \leq 1.5$ (Schmertmann 1978).

1. The ultimate bearing capacity of cohesionless soils is given by

$$\text{Strip:} \qquad q_u = 28 - 0.0052(300 - q_c)^{1.5} \tag{4-21a}$$

$$\text{Square:} \qquad q_u = 48 - 0.0090(300 - q_c)^{1.5} \tag{4-21b}$$

where q_u and q_c are in units of tsf or kg/cm^2.

2. The ultimate bearing capacity of cohesive soils is

$$\text{Strip:} \qquad q_u = 2 + 0.28q_c \tag{4-22a}$$

$$\text{Square:} \qquad q_u = 5 + 0.34q_c \tag{4-22b}$$

Units are also in tsf or kg/cm^2. Table 4-9 using Figure 4-3 provides a procedure for estimating q_u for footings up to $B = 8$ ft in width.

D. VANE SHEAR TEST. The vane shear is suitable for cohesive soil because bearing capacity is governed by short-term, undrained loading which is simulated by this test. Bearing capacity can be estimated by (Canadian Geotechnical Society 1985)

$$q_u = 5 \cdot R_v \cdot \tau_u\left[1 + 0.2 \cdot \frac{D}{B}\right]\left[1 + 0.2 \cdot \frac{B}{L}\right] + \sigma_{vo} \tag{4-24}$$

where

R_v = strength reduction factor, Figure 4-4

τ_u = field vane undrained shear strength measured during the test, ksf

D = depth of foundation, ft

Table 4-9. CPT procedure for estimating bearing capacity of footings on cohesive soil (data from Tand, Funegard, and Briaud 1986)

Step	Procedure
1.	Determine equivalent $\overline{q_c}$ from footing base to 1.5B below base by $$\overline{q_c} = (q_{cb1} \cdot q_{cb2})^{0.5} \quad (4\text{-}23a)$$ where $\overline{q_c}$ = equivalent cone tip bearing pressure below footing, ksf q_{cb1} = average tip resistance from 0.0 to 0.5B, ksf q_{cb2} = average cone tip resistance from 0.5B to 1.5B, ksf
2.	Determine equivalent depth of embedment D_e, ft, to account for effect of strong or weak soil above the bearing elevation $$D_e = \sum_{i=1}^{n} \Delta z_i \frac{q_{ci}}{q_c} \quad (4\text{-}23b)$$ where n = number of depth increments to depth D D = unadjusted (actual) depth of embedment, ft Δz_i = depth increment i, ft q_{ci} = cone tip resistance of depth increment i, ksf q_c = equivalent cone tip bearing pressure below footing, ksf
3.	Determine ratio of equivalent embedment depth to footing width $$R_d = \frac{D_e}{B} \quad (4\text{-}23c)$$
4.	Estimate bearing ratio R_k from Figure 4-3 using R_d. The lower bound curve is applicable to fissured or slickensided clays. The average curve is applicable to all other clays unless load tests verify the upper bound curve for intact clay.
5.	Estimate total overburden pressure σ_{vo}, then calculate $$q_{ua} = R_k(\overline{q_c} - \sigma_{vo}) + \sigma_{vo} \quad (4\text{-}23d)$$ where q_{ua} = ultimate bearing capacity of axially-loaded square or round footings with horizontal ground surface and base. Adjust q_{ua} obtained from Equation 4-23d for shape, eccentric loads, sloping ground or tilted base using Hansen's factors for cohesion, Table 4-5, to obtain the ultimate capacity $$q_a = \zeta_c \cdot q_{ua} \quad (4\text{-}23e)$$ where ζ_c is defined by Equation 4-8a.

B = width of foundation, ft
L = length of foundation, ft
σ_{vo} = total vertical soil overburden pressure at the foundation level, ksf

4-5. Examples

Estimation of the bearing capacity is given below for (**1**) a wall footing placed on the ground surface subjected to a vertical load, (**2**) a rectangular footing placed below the ground surface and subjected to an

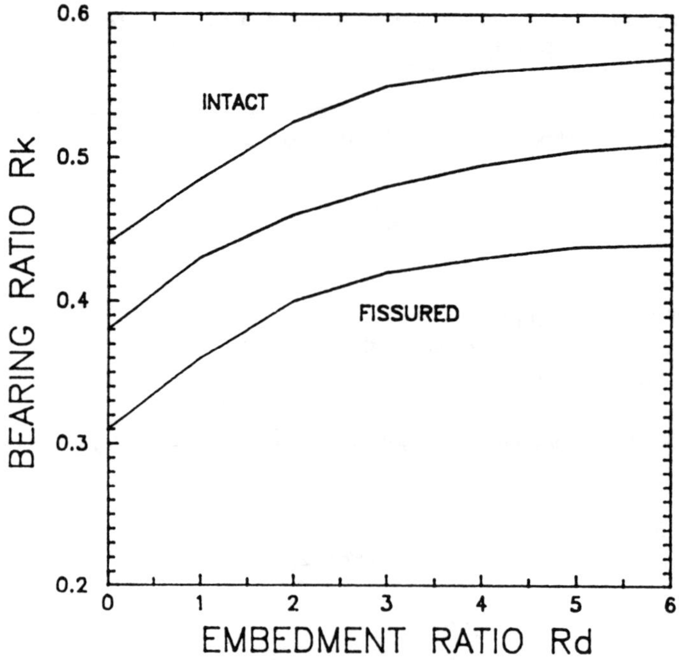

Figure 4-3. Bearing ratio R_k for axially-loaded square and round footings (Data from Tand, Funegard, and Briaud 1986)

inclined load, and (**3**) a tilted, rectangular footing on a slope and subjected to an eccentric load. Additional examples are provided in the user manual for CBEAR (Mosher and Pace 1982). The slope stability analysis of embankments is described in the user manual for UTEXAS2 (Edris 1987). Bearing capacity analyses should be performed using at least three methods where practical.

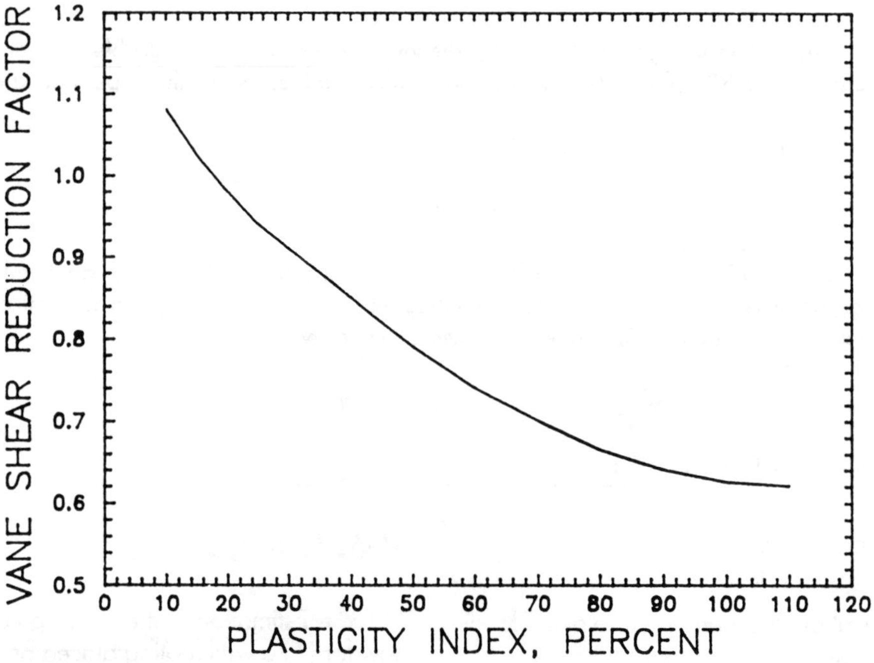

Figure 4-4. Strength reduction factor for field vane shear (Data from Bjerrum 1973)

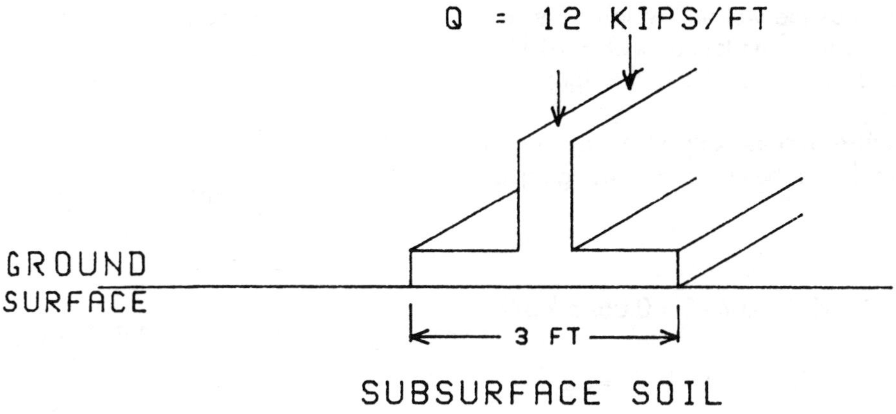

Figure 4-5. Example wall footing bearing capacity analysis

A. WALL FOOTING. A wall footing 3 ft wide with a load Q = 12 kips/ft (bearing pressure q = 4 ksf) is proposed to support a portion of a structure in a selected construction site. The footing is assumed to be placed on or near the ground surface for this analysis such that D = 0.0 ft, Figure 4-5, and $\sigma'_D = 0.0$. Depth H is expected to be < 2B or < 6 ft.

1. Soil Exploration. Soil exploration indicated a laterally uniform cohesive soil in the proposed site. Undrained triaxial compression test results were performed on specimens of undisturbed soil samples to determine the undrained shear strength. Confining pressures on these specimens were equal to the total vertical overburden pressure applied to these specimens when in the field. Results of these tests indicated the distribution of shear strength with depth shown in Figure 4-6. The minimum shear strength $c = C_u$ of 1.4 ksf observed 5 to 7 ft below ground surface is selected for the analysis. The friction angle is ϕ = 0.0 deg and the wet unit weight is 120 psf.

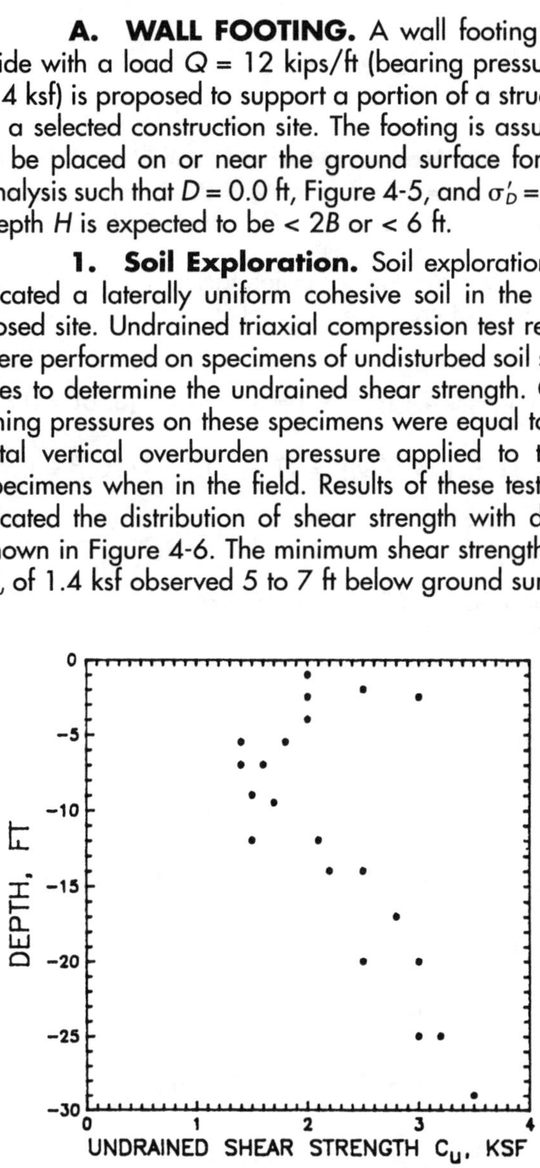

Figure 4-6. Example undrained shear strength distribution with depth

2. Ultimate Bearing Capacity

a. Terzaghi Method. Table 4-1 indicates N_c = 5.7, N_q = 1.0 and N_γ = 0.00. The total ultimate capacity q_u is

$$q_u = cN_c = 1.4 \cdot 5.7 = 8.0 \text{ ksf}$$

The Terzaghi method indicates an ultimate bearing capacity q_u = 8 ksf.

b. Meyerhof Method. The ultimate bearing capacity of this wall footing using program CBEAR yields q_u = 7.196 ksf. The Hansen and Vesic solutions will be similar.

3. Allowable Bearing Capacity. *FS* for this problem from Table 1-2 is 3.0. Therefore, q_a using Equation 1-2a is q_u/FS = 8.000/3 = 2.7 ksf from the Terzaghi solution and 7.196/3 = 2.4 ksf from CBEAR. The solution using program UTEXAS2 gives a minimum *FS* = 2.2 for a circular failure surface of radius 3 ft with its center at the left edge of the footing.

4. Recommendation. q_a ranges from 2.4 to 2.7 while the proposed design pressure q_d is 4 ksf. q_d should be reduced to 2.4 ksf $\leq q_a$.

B. RECTANGULAR FOOTING WITH INCLINED LOAD. A rectangular footing with B = 3 ft, W = 6 ft, D = 2 ft, similar to Figure 1-6, is to be placed in cohesionless soil on a horizontal surface (β = 0.0) and without base tilt (δ = 0.0). The effective friction angle ϕ' = 30 deg and cohesion $c = c_u$ = 0.0. The surcharge soil has a wet (moist) unit weight γ_D = 0.120 kip/ft^3 (120 pcf), subsurface soil has a wet (moist) unit weight γ_H = 0.130 kip/ft^3 (130 pcf), and depth to groundwater is D_{GWT} = 3 ft. The saturated unit weight is assumed the same as the wet unit weight. The applied vertical load on the foundation is Q = 10 kips and the horizontal load T = +2 kips to the right.

1. Effective Stress Adjustment. Adjust the unit soil weights due to the water table using Equation 1-6

$$\gamma_{HSUB} = \gamma_H - \gamma_w = 0.130 - 0.0625 = 0.0675 \text{ kip/ft}^3$$

$$H = B \cdot \tan(45 + \phi/2) = 3.00 \cdot 1.73 = 5.2 \text{ ft}$$

$$\gamma'_H = \gamma_{HSUB} + [(D_{GWT} - D)/H] \cdot \gamma_w$$

$$= 0.0675 + [(3 - 2)/5.2] \cdot 0.0625$$

$$= 0.08 \text{ kip/ft}^3$$

From Equation 1-7a, $\sigma'_D = \sigma_D = \gamma_D \cdot D = 0.120 \cdot 2.00 = 0.24$ ksf.

2. Meyerhof Method. For ϕ' = 30 deg, N_q = 18.40, N_γ = 15.67, and N_ϕ = 3.00 from Table 4-4. N_c is not needed since c = 0.0. From Table 4-3,

a. Wedge correction factor $\zeta_\gamma = \zeta_{\gamma s} \cdot \zeta_{\gamma i} \cdot \zeta_{\gamma d}$

$$\zeta_{\gamma s} = 1 + 0.1 \cdot N_\phi \cdot (B'/W')$$

$$= 1 + 0.1 \cdot 3.00 \cdot (3/6) = 1.15$$

$$R = (Q^2 + T^{\ \ 2})^{0.5}$$

$$= (100 + 4)^{0.5} = 10.2$$

$$\theta = \cos^{-1}(Q/R) = \cos^{-1}(10/10.2)$$

$$= 11.4 \text{ deg} < \phi = 30 \text{ deg}$$

$$\zeta_{\gamma i} = [1 - (\theta/\phi')]^2 = [1$$

$$- (11.4/30)]^2 = 0.384$$

$$\zeta_{\gamma d} = 1 + 0.1\sqrt{N_\phi} \cdot (D/B)$$

$$= 1 + 0.1 \cdot 1.73 \cdot (2/3) = 1.115$$

$$\zeta = 1.15 \cdot 0.384 \cdot 1.115 = 0.49$$

b. Surcharge correction factor $\zeta_q = \zeta_{qs} \cdot \zeta_{qi} \cdot \zeta_{qd}$

$$\zeta_{qs} = \zeta_{\gamma s} = 1.15$$

$$\zeta_{qi} = [1 - (\theta/90)]^2 = [1 - (11.4/90)]^2 = 0.763$$

$$\zeta_{qd} = \zeta_{\gamma d} = 1.115$$

$$\zeta_q = 1.15 \cdot 0.763 \cdot 1.115 = 0.98$$

c. Total ultimate bearing capacity from Equation 4-1 is

$$q_u = 0.5 \cdot B \cdot \gamma'_H N_\gamma \cdot \zeta_\gamma \quad + \sigma'_D \cdot N_q \cdot \zeta_q$$

3. Hansen Method. For ϕ' = 30 deg, N_q = 18.40, N_γ = 15.07, and N_ϕ = 3.00 from Table 4-4. N_c is not needed since c = 0.0. From Table 4-5,

a. Wedge correction factor $\zeta_\gamma = \zeta_{\gamma s} \cdot \zeta_{\gamma i} \cdot \zeta_{\gamma d} \cdot \zeta_{\gamma\beta} \cdot \zeta_{\gamma\delta}$ where $\zeta_{\gamma\beta} = \zeta_{\gamma\delta} = 1.00$

$$\zeta_{\gamma s} = 1 - 0.4 \cdot (B'/W') = 1 - 0.4 \cdot (3/6) = 0.80$$

$$\zeta_{\gamma i} = [1 - (0.7T/Q)]^5 = [1 - (0.7 \cdot T/10)]^5 = 0.47$$

$$\zeta_{\gamma d} = 1.00$$

$$\zeta_\gamma = 0.80 \cdot 0.47 \cdot 1.00 = 0.376$$

b. Surcharge correction factor $\zeta_q = \zeta_{qs} \cdot \zeta_{qi} \cdot \zeta_{qd} \cdot \zeta_{q\beta} \cdot \zeta_{q\delta}$ where $\zeta_{q\beta} = \zeta_{q\delta} = 1.00$

$$\zeta_{qs} = 1 + (B/W) \cdot \tan\phi = 1 + (3/6) \cdot 0.577 = 1.289$$

$$\zeta_{qi} = [1 - (0.5T/Q)]^5 = [1 - (0.5 \cdot 2/10)]^5 = 0.59$$

$$k = D/B = 2/3$$

$$\zeta_{qd} = 1 + 2 \cdot \tan\phi' \cdot (1 - \sin\phi')^2 \cdot k$$

$$= 1 + 2 \cdot 0.577 \cdot (1 - 0.5)^2 \cdot 2/3$$

$$= 1.192$$

$$\zeta_q = 1.289 \cdot 0.59 \cdot 1.192 = 0.907$$

c. Total ultimate bearing capacity from Equation 4-1 is

$$q_u = 0.5 \cdot B \cdot \gamma'_H \cdot N_\gamma \cdot \zeta_\gamma + \sigma'_D \cdot N_q \cdot \zeta_q$$

$$= 0.5 \cdot 3.00 \cdot 0.08 \cdot 15.07 \cdot 0.376 + 0.24 \cdot 18.40 \cdot 0.907$$

$$= 0.68 + 4.01 = 4.69 \text{ ksf}$$

4. Vesic Method. For $\phi' = 30$ deg, $N_q = 18.40$, $N_\gamma = 22.40$, and $N_\phi = 3.00$ from Table 4-4. N_c is not needed. From Table 4-6,

a. Wedge correction factor $\zeta_\gamma = \zeta_{\gamma s} \cdot \zeta_{\gamma i} \cdot \zeta_{\gamma d} \cdot \zeta_{\gamma \beta} \cdot \zeta_{\gamma \delta}$ where $\zeta_{\gamma\beta} = \zeta_{\gamma\delta} = 1.00$

$$\zeta_{\gamma s} = 1 - 0.4 \cdot B/W = 1 - 0.4 \cdot 3/6 = 0.80$$

$$R_{BW} = B/W = 3/6 = 0.5$$

$$m = (2 + R_{BW})/(1 + R_{BW}) = (2 + 0.5)/(1 + 0.5) = 1.67$$

$$\zeta_{\gamma i} = [(1 - (T/Q)]^{m+1} = [1 - (2/10)]^{1.67+1} = 0.551$$

$$\zeta_\gamma = 0.80 \cdot 0.551 \cdot 1.00 = 0.441$$

b. Surcharge correction factor $\zeta_q = \zeta_{qs} \cdot \zeta_{qi} \cdot \zeta_{qd} \cdot \zeta_{q\beta} \cdot \zeta_{q\delta}$ where $\zeta_{q\beta} = \zeta_{q\delta} = 1.00$

$$\zeta_{qs} = 1 + (B/W) \cdot \tan\phi = 1 + 3/6 \cdot 0.577 = 1.289$$

$$\zeta_{qi} = [1 - (T/Q)]^m = [1 - (2/10)]^m = 0.689$$

$$\zeta_{qd} = 1 + 2 \cdot \tan\phi' \cdot (1 - \sin\phi')^2 \cdot k$$

$$p = 1 + 2 \cdot 0.577 \cdot (1 - 0.5) \cdot 2/3$$

$$= 1.192$$

$$\zeta_q = 1.289 \cdot 0.689 \cdot 1.192 = 1.058$$

c. Total ultimate bearing capacity from Equation 3-1a is

$$q_u = 0.5 \cdot B \cdot \gamma'_H \cdot N_\gamma \cdot \zeta_\gamma + \sigma'_D \cdot N_q \cdot \zeta_q$$

$$= 0.5 \cdot 3.00 \cdot 0.08 \cdot 22.40 \cdot 0.441 + 0.24 \cdot 18.40 \cdot 1.058$$

$$= 1.19 + 4.67 = 5.86 \text{ ksf}$$

5. Program CBEAR. Zero elevation for this problem is defined 3 ft below the foundation base. Input to this program is as follows (refer to Figure 1-6):

a. Foundation coordinates:

$$x_1 = 10.00, \quad y_1 = 3.00$$

$$x_2 = 13.00, \quad y_2 = 3.00$$

Length of footing: = 6.00

b. Soil Coordinates:

$$x_1 = x_{s1} = 10.00, \; y_1 = y_{s1} = 3.00$$

(top elevation of $x_2 = x_{s2} = 13.00$, $y_2 = y_{s2} = 3.00$

subsurface soil)

c. Soil Properties:

moist (wet) unit weight γ_H = 130 pounds/ft³

(subsurface soil) saturated unit weight = γ_H

friction angle = 30 deg

cohesion = 0.00

d. Options:

9One surcharge	*y* coordinate of top of surcharge	= 5.00 ft
layer	moist unit weight	= 120 pounds/ft³
	saturated unit weight	= 120 pounds/ft³
Water table	*y* coordinate of top of water table	= 2.00 ft
description	unit weight of water	= 62.5 pounds/ft³
Applied load	applied load (*R*)	= 10.2 kips
description	*x* coordinate of base application point	= 11.5 ft
	z coordinate of base application point	= 3.00 ft
	inclination of load clockwise from vertical	= 11.4 deg

e. CBEAR calculates $q_u = 5.34$ ksf

f. Comparison of methods indicates bearing capacities

Method	Total q_u, ksf	Net q'_u, ksf
Meyerhof	5.25	5.01
Hansen	4.69	4.45
Vesic	5.86	5.62
Program CBEAR	5.34	5.10

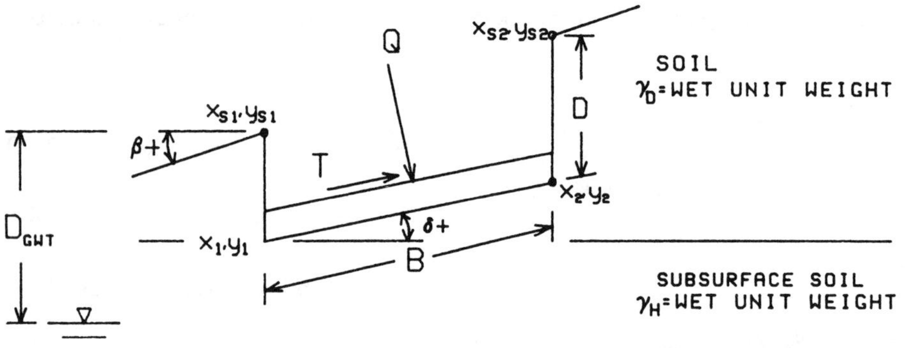

Figure 4-7. Shallow foundation with slope and base tilt

The net bearing capacity is found by subtracting $\gamma_D \cdot D = 0.12 \cdot 2 = 0.24$ ksf from q_u, Equation 4-2. The resultant applied pressure on the footing is $q_r = R/(BW) = 10.2/(3 \cdot 6) = 0.57$ ksf. The factor of safety of all of the above methods with respect to the net bearing capacity is on the order of $q_u'/q_r \approx 9$. The Hansen method is most conservative.

C. RECTANGULAR FOOTING WITH ECCENTRICITY, BASE TILT, AND GROUND SLOPE. A rectangular footing, $B = 3$ ft and $W = 5$ ft, is placed in a cohesionless soil with base tilt $\delta = 5$ deg and ground slope $\beta = 15$ deg as illustrated in Table 4-5 and Figure 4-7. $\phi' = 26$ deg and $c = c_a = 0.0$. Soil wet unit weight $\gamma_D = 120$ lbs/ft^3, subsurface soil wet unit weight $\gamma_H = 130$ lbs/ft^3, and depth to groundwater $D_{GWT} = 3$ ft. Vertical applied load $Q = 10$ kips and horizontal load $T = 0$ kips, but $M_B = 5$ kips-ft and $M_W = 10$ kips-ft.

1. Coordinate Adjustment. $\delta = 5$ deg indicates right side elevation of the base is $3 \cdot \sin 5$ deg $= 0.26$ ft higher than the left side. $\beta = 15$ deg indicates right side foundation elevation at the ground surface is $3 \cdot \sin 15$ deg $= 0.78$ ft higher than the left side.

2. Effective Stress Adjustment. Average $D_{GWT} = 3 + 0.78/2 = 3.39$ ft. Average $D = 2 + 0.78/2 - 0.26/2 = 2.26$ ft. Adjustment of soil unit wet weight for the water table from Equation 1-6 is

$$\gamma_{HSUB} = \gamma_H - \gamma_w = 0.130 - 0.0625 = 0.0675 \text{ kip/ft}^3$$

$$H = B \cdot \tan[45 + (\phi/2)] = 3.00 \cdot 1.73 = 5.2 \text{ ft}$$

$$\gamma_H' = \gamma_{HSUB} + [(D_{GWT} - D)/H] \cdot \gamma_w$$

$$= 0.0675 + [(3.39 - 2.26)/5.2] \cdot 0.0625 = 0.081 \text{ kip/ft}^3$$

$$\sigma_D' = \sigma_D = \gamma_D \cdot D = 0.120 \cdot 2.26 = 0.27 \text{ ksf}$$

3. Eccentricity Adjustment. Bending moments lead to eccentricities from Equations 4-4c and 4-4d

$$e_B = M_B/Q = 5/10 = 0.5 \text{ ft}$$

$$e_W = M_W/Q = 10/10 = 1.0 \text{ ft}$$

Effective dimensions from Equations 4-4a and 4-4b are

$$B' = B - 2e_B = 3 - 2 \cdot 0.5 = 2 \text{ ft}$$

$$W' = W - 2e_W = 5 - 2 \cdot 1.0 = 3 \text{ ft}$$

4. Hansen Method. For $\phi' = 26$ deg, $N_q = 11.85$ and $N_\gamma = 7.94$ from Table 4-4. N_c is not needed since $c = 0.0$. From Table 4-5,

a. Wedge correction factor $\zeta_\gamma = \zeta_{\gamma s} \cdot \zeta_{\gamma i} \cdot \zeta_{\gamma d} \cdot \zeta_{\gamma\beta} \cdot \zeta_{\gamma\delta}$ where $\zeta_{\gamma i} = 1.00$

$$\zeta_{\gamma s} = 1 - 0.4 \cdot B'/W' = 1 - 0.4 \cdot 2/3 = 0.733$$

$$\zeta_{\gamma d} = 1.00$$

$$\zeta_{\gamma\beta} = (1 - 0.5 \cdot \tan\beta)^5 = (1 - 0.5 \cdot \tan 15)^5 = 0.487$$

$$\zeta_{\gamma\delta} = e^{-0.047 \cdot \delta \cdot \tan\phi'} = e^{-0.047 \cdot 5 \cdot \tan 26} = 0.892$$

$$\zeta_\gamma = 0.733 \cdot 1.000 \cdot 0.487 \cdot 0.892 = 0.318$$

b. Surcharge correction factor $\zeta_q = \zeta_{qs} \cdot \zeta_{qi} \cdot \zeta_{qd} \cdot \zeta_{q\beta} \cdot \zeta_{q\delta}$ where $\zeta_{qi} = 1.00$

$$\zeta_{qs} = 1 + (B'/W') \cdot \tan\phi = 1 + (2/3) \cdot 0.488 = 1.325$$

$$k = D/B = 2.26/3 = 0.753$$

$$\zeta_{qd} = 1 + 2 \cdot \tan\phi' \cdot (1 - \sin\phi')^2 \cdot k = 1 + 2 \cdot 0.488 \cdot (1 - 0.438) \cdot 0.753$$

$$\zeta_{qd} = 1.232$$

$$\zeta_{q\beta} = \zeta_{\gamma\beta} = 0.487$$

$$\zeta_{q\delta} = e^{-0.035 \cdot \delta \cdot \tan\phi'} = e^{-0.035 \cdot 5 \cdot \tan 26} = 0.918$$

$$\zeta_q = 1.325 \cdot 1.232 \cdot 0.487 \cdot 0.918 = 0.730$$

c. Total ultimate bearing capacity from Equation 4-1 is

$$q_u = 0.5 \cdot B \cdot \gamma'_H \cdot N_\gamma \cdot \zeta_\gamma + \sigma'_D \cdot N_q \cdot \zeta_q$$

$$= 0.5 \cdot 2.00 \cdot 0.081 \cdot 7.942 \cdot 0.318 + 0.27 \cdot 11.85 \cdot 0.730$$
$$= 0.205 + 2.335 = 2.54 \text{ ksf}$$

5. Vesic Method. For $\phi' = 26$ deg, $N_q = 11.85$ and $N_\gamma = 12.54$ from Table 4-4. N_c is not needed. From Table 4-6,

a. Wedge correction factor $\zeta_\gamma = \zeta_{\gamma s} \cdot \zeta_{\gamma i} \cdot \zeta_{\gamma d} \cdot \zeta_{\gamma\beta} \cdot \zeta_{\gamma\delta}$ where $\zeta_{\gamma i} = 1.00$

$$\zeta_{\gamma s} = 1 - 0.4 \cdot B/W = 1 - 0.4 \cdot 2/3 = 0.733$$

$$\zeta_{\gamma d} = 1.00$$

$$\zeta_{\gamma\beta} = (1 - \tan\beta)^2 = (1 - \tan 15)^2 = 0.536$$

$$\zeta_{\gamma\delta} = (1 - 0.017 \cdot \tan\phi')^2 = (1 - 0.017 \cdot \delta \cdot 5 \cdot \tan 26)^2 = 0.919$$

$$\zeta_\gamma = 0.733 \cdot 1.00 \cdot 0.536 \cdot 0.919 = 0.361$$

b. Surcharge correction factor $\zeta_q = \zeta_{qs} \cdot \zeta_{qi} \cdot \zeta_{qd} \cdot \zeta_{q\beta} \cdot \zeta_{q\delta}$ where $\zeta_{qi} = \zeta_{q\delta} = 1.00$

$$\zeta_{qs} = 1 + (B/W) \cdot \tan\phi = 1 + 2/3 \cdot 0.488 = 1.325$$

$$\zeta_{qd} = 1 + 2 \cdot \tan\phi' \cdot (1 - \sin\phi')^2 \cdot k$$

$$\zeta_{qd} = 1 + 2 \cdot 0.488 \cdot (1 - 0.438)^2 \cdot 0.753 = 1.232$$

$$\zeta_{q\beta} = \zeta_{\gamma\beta} = 0.536$$

$$\zeta_{q\delta} = \zeta_{\gamma\delta} = 0.919$$

$$\zeta_q = 1.325 \cdot 1.232 \cdot 0.536 \cdot 0.919 = 0.804$$

c. Total ultimate bearing capacity from Equation 4-1 is

$$q_u = 0.5 \cdot B \cdot \gamma'_H \cdot N_\gamma \cdot \zeta_\gamma + \sigma'_D \cdot N_q \cdot \zeta_q$$

$$= 0.5 \cdot 2.00 \cdot 0.081 \cdot 12.54 \cdot 0.361 + 0.27 \cdot 11.85 \cdot 0.804$$
$$= 0.367 + 2.572 = 2.94 \text{ ksf}$$

6. Program CBEAR. Input is as follows (refer to Figure 4-7):

a. Foundaation coordinates:

$$x_1 = 10.00, \quad y_1 = 3.00$$

$$x_2 = 13.00, \quad y_2 = 3.26$$

Length of footing: = 5.00

b. Soil Coordinates:

$$x_{s1} = 10.00, \quad y_{s1} = 5.00$$

$$x_{s2} = 13.00, \quad y_{s2} = 5.78$$

c. SoilProperties:

moist (wet) unit weight	γ_H = 120 pounds/ft³
saturated unit weight	= γ_H
friction angle	= 26 deg
cohesion	= 0.00

d. Options:

One surcharge layer	*y* coordinate of top of subsurface soil	= 3.00 ft
	moist unit weight	= 130 pounds/ft³
	saturated unit weight	= 130 pounds/ft³
	friction angle	= 26 degrees
	cohesion	= 0.0
Water table description	*y* coordinate of top of water table	= 2.00 ft
	unit weight of water	= 62.5 pounds/ft³
Applied load description	applied load (*R*)	= 10.0 kips
	x coordinate of base application point	= 11.0 ft
	z coordinate of base application point	= 2.00 ft
	inclination of load clockwise from vertical	= 0.0 deg

e. CBEAR calculates $q_u = 2.21$ ksf

f. Comparison of methods indicates bearing capacities

Method	Total q_u, ksf	Net q_u', ksf
Hansen	2.55	2.28
Vesic	3.94	2.67
Program CBEAR	2.21	1.94

Net bearing capacity is found by subtracting $\gamma_D \cdot D = 0.12 \cdot (2 + 2.78)/2 = 0.27$ ksf from q_u, Equation 4-2. The resultant applied pressure on the footing is $q_r = Q/(B'W') = 10/(2 \cdot 3) = 1.67$ ksf. The factors of safety of all of the above methods are $q_u'/q_r < 2$. The footing is too small for the applied load and bending moments. Program CBEAR is most conservative. CBEAR ignores subsoil data if the soil is sloping and calculates bearing capacity for the footing on the soil layer only.

CHAPTER 5

DEEP FOUNDATIONS

5-1. Basic Considerations

Deep foundations transfer loads from structures to acceptable bearing strata at some distance below the ground surface. These foundations are used when the required bearing capacity of shallow foundations cannot be obtained, settlement of shallow foundations is excessive, and shallow foundations are not economical. Deep foundations are also used to anchor structures against uplift forces and to assist in resisting lateral and overturning forces. Deep foundations may also be required for special situations such as expansive or collapsible soil and soil subject to erosion or scour.

A. DESCRIPTION. Bearing capacity analyses are performed to determine the diameter or cross-section, length, and number of drilled shafts or driven piles required to support the structure.

1. Drilled Shafts. Drilled shafts are nondisplacement reinforced concrete deep foundation elements constructed in dry, cased, or slurry-filled boreholes. A properly constructed drilled shaft will not cause any heave or loss of ground near the shaft and will minimize vibration and soil disturbance. Dry holes may often be bored within 30 minutes leading to a rapidly constructed, economical foundation. Single-drilled shafts may be built with large diameters and can extend to deep depths to support large loads. Analysis of the bearing capacity of drilled shafts is given in Section I.

a. Lateral expansion and rebound of adjacent soil into the bored hole may decrease pore pressures. Heavily overconsolidated clays and shales may weaken and transfer some load to the shaft base where pore pressures may be positive. Methods presented in Section I for calculating bearing capacity in clays may be slightly unconservative, but the FS's should provide an adequate margin of safety against overload.

b. Rebound of soil at the bottom of the excavation and water collecting at the bottom of an open bore hole may reduce end bearing capacity and may require construction using slurry.

c. Drilled shafts tend to be preferred to driven piles as the soil becomes harder, pile driving becomes difficult, and driving vibrations affect nearby structures. Good information concerning rock is required when drilled shafts are carried to rock. Rock that is more weathered or of lesser quality than expected may require shaft bases to be placed deeper than expected. Cost overruns can be significant unless good information is available.

2. Driven Piles. Driven piles are displacement deep foundation elements driven into the ground causing the soil to be displaced and disturbed or remolded. Driving often temporarily increases pore pressures and reduces short term bearing capacity, but may increase long-term bearing capacity. Driven piles are often constructed in groups to provide adequate bearing capacity. Analysis of the bearing capacity of driven piles and groups of driven piles is given in Section II.

a. Driven piles are frequently used to support hydraulic structures such as locks and retaining walls and to support bridges and highway overpasses. Piles are also useful in flood areas with unreliable soils.

b. Pile driving causes vibration with considerable noise and may interfere with the performance of nearby structures and operations. A preconstruction survey of nearby structures may be required.

c. The cross-section and length of individual piles are restricted by the capacity of equipment to drive piles into the ground.

d. Driven piles tend to densify cohesionless soils and may cause settlement of the surface, particularly if the soil is loose.

e. Heave may occur at the surface when piles are driven into clay, but a net settlement may occur over the long term. Soil heave will be greater in the direction toward which piles are placed and driven. The lateral extent of ground heave is approximately equal to the depth of the bottom of the clay layer.

3. Structural capacity. Stresses applied to deep foundations during driving or by structural loads should be compared with the allowable stresses of materials carrying the load.

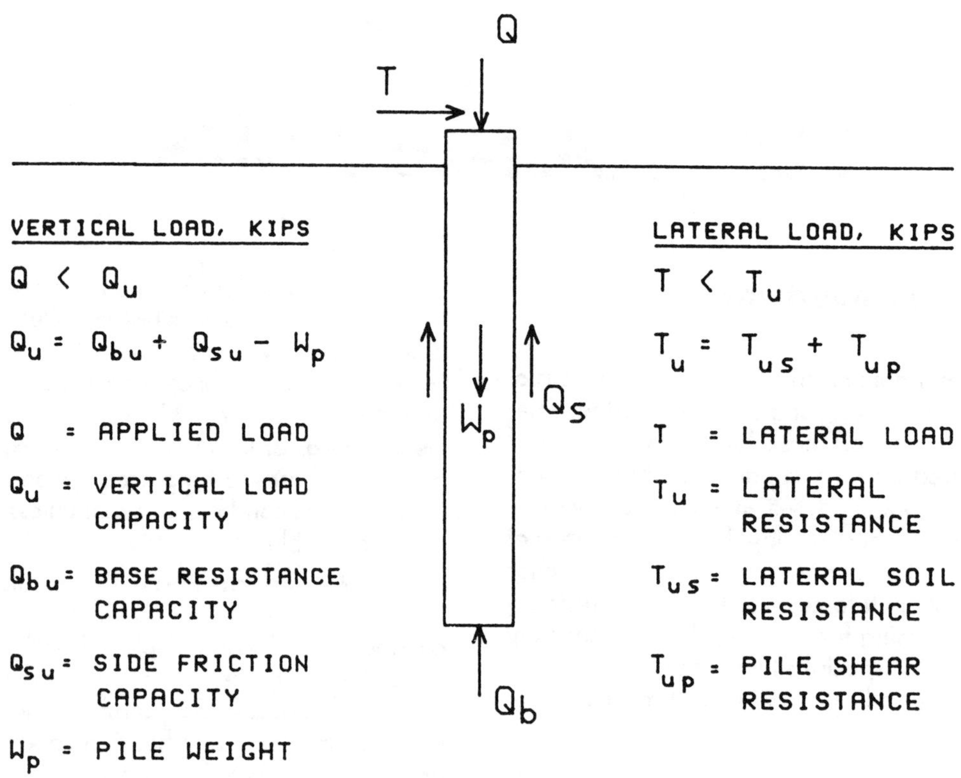

Figure 5-1. Support of deep foundations

B. DESIGN RESPONSIBILITY. Selection of appropriate design and construction methods requires geotechnical and structural engineering skills. Knowledge of how a deep foundation interacts with the superstructure is provided by the structural engineer with soil response information provided by the geotechnical engineer. Useful soil-structure interaction analyses can then be performed of the pile-soil support system.

C. LOAD CONDITIONS. Mechanisms of load transfer from the deep foundation to the soil are not well understood and complicate the analysis of deep foundations. Methods available and presented below for evaluating ultimate bearing capacity are approximate. Consequently, load tests are routinely performed for most projects, large or small, to determine actual bearing capacity and to evaluate performance. Load tests are not usually performed on drilled shafts carried to bedrock because of the large required loads and high cost.

1. Representation of Loads. The applied loads may be separated into vertical and horizontal components that can be evaluated by soil-structure interaction analyses and computer-aided methods. Deep foundations must be designed and constructed to resist both applied vertical and lateral loads (Figure 5-1). The applied vertical load Q is supported by soil-shaft side friction Q_{su} and base resistance Q_{bu}. The applied lateral load T is carried by the adjacent lateral soil and structural resistance of the pile or drilled shaft in bending (Figure 5-2).

a. Applied loads should be sufficiently less than the ultimate bearing capacity to avoid excessive vertical and lateral displacements of the pile or drilled shaft. Displacements should be limited to 1 inch or less.

b. Factors of safety applied to the ultimate bearing capacity to obtain allowable loads are often 2 to 4. *FS* applied to estimations of the ultimate bearing capacity from static load test results should be 2.0. Otherwise, *FS* should be at least 3.0 for deep foundations in both clay and sand. *FS* should be 4 for deep foundations in multi-layer clay soils and clay with undrained shear strength $C_u > 6$ ksf.

2. Side Friction. Development of soil-shaft side friction resisting vertical loads leads to relative movements between the soil and shaft. The maximum side friction is often developed after relative small displacements less than 0.5 inch. Side friction is limited by the adhesion between the shaft and the soil or else the shear strength of the adjacent soil, whichever is smaller.

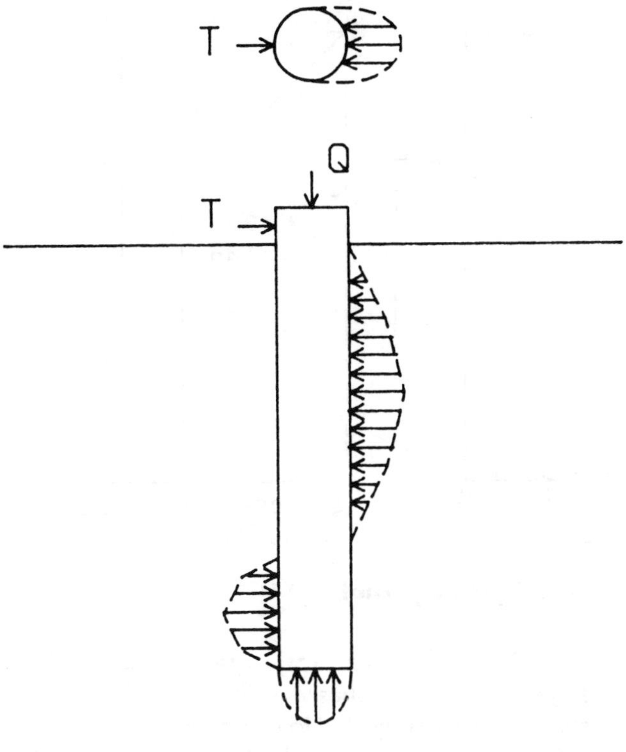

Figure 5-2. Earth pressure distribution T_{us} acting on a laterally-loaded pile

a. Side friction often contributes the most bearing capacity in practical situations unless the base is bearing on stiff shale or rock that is much stiffer and stronger than the overlying soil.

b. Side friction is hard to accurately estimate, especially for foundations constructed in augered or partially jetted holes or foundations in stiff, fissured clays.

3. Base Resistance. Failure in end bearing normally consists of a punching shear at the tip. Applied vertical compressive loads may also lead to several inches of compression prior to a complete plunging failure. The full soil shear strength may not be mobilized beneath the pile tip and a well-defined failure load may not be observed when compression is significant.

SECTION I. DRILLED SHAFTS

5-2. Vertical Compressive Capacity of Single Shafts

The approximate static load capacity of single drilled shafts from vertical applied compressive forces is

$$Q_u \approx Q_{bu} + Q_{su} - W_p \quad (5\text{-}1a)$$

$$Q_u \approx q_{bu}A_b + \sum_{i=1}^{n} Q_{sui} - W_p \quad (5\text{-}1b)$$

where

Q_u = ultimate drilled shaft or pile resistance, kips
Q_{bu} = ultimate end bearing resistance, kips
Q_{su} = ultimate skin friction, kips
q_{bu} = unit ultimate end bearing resistance, ksf
A_b = area of tip or base, ft^2
n = number of increments the pile is divided for analysis (referred to as a pile element, Figure C-1)
Q_{sui} = ultimate skin friction of pile element i, kips
W_p = pile weight, $\approx A_b \cdot L \cdot \gamma_p$ without enlarged base, kips
L = pile length, ft
γ_p = pile density, kips/ft^3

A pile may be visualized to consist of a number of elements as illustrated in Figure C-1, Appendix C, for the calculation of ultimate bearing capacity.

A. END BEARING CAPACITY. Ultimate end bearing resistance at the tip may be given as Equation 4-1 neglecting pile weight W_p

$$q_{bu} = c \cdot N_{cp} \cdot \zeta_{cp} + \sigma'_L \cdot N_{qp}\zeta_{qp} + \frac{B_b}{2} \cdot \gamma'_b \cdot N_{\gamma p}\,\zeta_{\gamma p} \quad (5\text{-}2a)$$

where

c = cohesion of soil beneath the tip, ksf
σ'_L = effective soil vertical overburden pressure at pile base $\approx \gamma'_L \cdot L$, ksf
γ'_L = effective wet unit weight of soil along shaft length L, kips/ft^3
B_b = base diameter, ft
γ'_b = effective wet unit weight of soil in failure zone beneath base, kips/ft^3
$N_{cp}, N_{qp}, N_{\gamma p}$ = pile bearing capacity factors of cohesion, surcharge, and wedge components

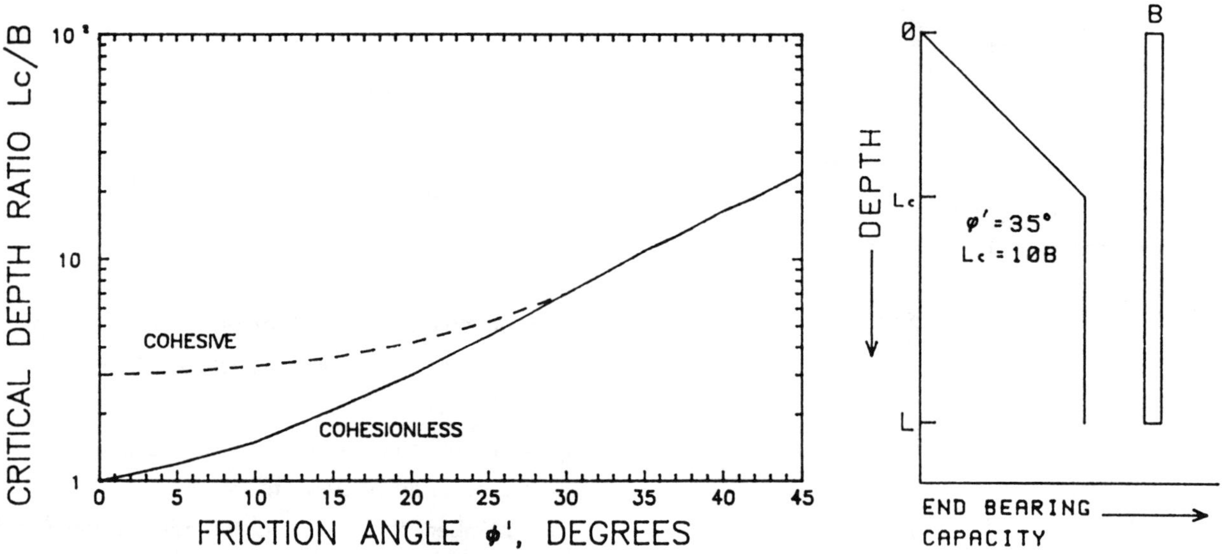

Figure 5-3. Critical depth ratio *Lc/B* (Data from Meyerhof 1976)

$\zeta_{cp}, \zeta_{qp}, \zeta_{\gamma p}$ = pile soil and geometry correction factors of cohesion, surcharge, and wedge components

Methods for estimating end bearing capacity and correction factors of Equation 5-2a should consider that the bearing capacity reaches a limiting constant value after reaching a certain critical depth. Methods for estimating end bearing capacity from in situ tests are discussed in Section II on driven piles.

1. Critical Depth. The effective vertical stress appears to become constant after some limiting or critical depth *Lc*, perhaps from arching of soil adjacent to the shaft length. The critical depth ratio *Lc/B* where *B* is the shaft diameter may be found from Figure 5-3. The critical depth applies to the Meyerhof and Nordlund methods for analysis of bearing capacity.

2. Straight Shafts. Equation 5-2a may be simplified for deep foundations without enlarged tips by eliminating the N_p term

$$q_{bu} = c \cdot N_{cp} \cdot \zeta_{cp} + \sigma'_L \cdot (N_{qp} - 1) \cdot \zeta_{qp} \quad (5\text{-}2b)$$

or

$$q_{bu} = c \cdot N_{cp} \cdot \zeta_{cp} + \sigma'_L \cdot N_{qp} \cdot \zeta_{qp} \quad (5\text{-}2c)$$

Equations 5-2b and 5-2c also compensates for pile weight W_p assuming $\gamma_p \approx \gamma'_L$. Equation 5-2c is usually used rather than Equation 5-2b because N_{qp} is usually large compared with "1" and $N_{qp} - 1 \approx N_{qp}$. W_p in Equation 5-1 may be ignored when calculating Q_u.

3. Cohesive Soil. The undrained shear strength of saturated cohesive soil for deep foundations in saturated clay subjected to a rapidly applied load is $c = C_u$ and the friction angle $\phi = 0$. Equations 5-2 simplifies to (Reese and O'Neill 1988)

$$q_{bu} = F_r N_{cp} C_u, \; q_{bu} \leq 80 \text{ ksf} \quad (5\text{-}3)$$

where the shape factor $\zeta_{cp} = 1$ and $N_{cp} = 6 \cdot [1 + 0.2(L/B_b)] \leq 9$. The limiting q_{bu} of 80 ksf is the largest value that has so far been measured for clays. C_u may be reduced by about 1/3 in cases where the clay at the base has been softened and could cause local high strain bearing failure. F_r should be 1.0, except when B_b exceeds about 6 ft. For base diameter $B_b > 6$ ft,

$$F_r = \frac{2.5}{aB_b + 2.5b}, \; F_r \leq 1.0 \quad (5\text{-}4)$$

where

$a = 0.0852 + 0.0252(L/B_b), \; a \leq 0.18$

$b = 0.45C_u^{0.5}, \; 0.5 \leq b \leq 1.5$

The undrained strength of soil beneath the base C_u is in units of ksf. Equation 5-3 limits q_{bu} to bearing pressures for a base settlement of 2.5 inches. The undrained shear strength C_u is estimated by methods in Chapter 3 and may be taken as the average shear strength within $2B_b$ beneath the tip of the shaft.

4. Cohesionless Soil. Hanson, Vesic, Vesic Alternate, and general shear methods of estimating the bearing capacity and adjustment factors are recommended for solution of ultimate end bearing capacity using Equations 5-2. The Vesic method requires volumetric strain data ϵ_v of the foundation soil in addition to the effective friction angle ϕ'. The Vesic Alternate method provides a lower bound estimate of bearing capacity. The Vesic Alternate method may also be more appropriate for deep foundations constructed under difficult conditions, for drilled shafts placed in soil subject to disturbance and when a bentonite-water slurry is used to keep the hole open during drilled shaft construction. Several of these methods should be used for each design problem to provide a reasonable range of the probable bearing capacity if calculations indicate a significant difference between methods.

a. Hanson Method. The bearing capacity factors N_{cp}, N_{qp}, and $N_{\gamma p}$ and correction factors ζ_{cp}, ζ_{qp}, and $\zeta_{\gamma p}$ for shape and depth from Table 4-5 may be used to evaluate end bearing capacity using Equations 5-2. Depth factors ζ_{cd} and ζ_{qd} contain a "k" term that prevents unlimited increase in bearing capacity with depth. $k = \tan^{-1}(L_b/B)$ in radians where L_b is the embedment depth in bearing soil and B is the shaft diameter. $L_b/B \leq L_c/B$, Figure 5-3.

b. Vesic Method. The bearing capacity factors of Equation 5-2b are estimated by (Vesic 1977)

$$N_{cp} = (N_{qp} - 1) \cdot \cot \phi' \qquad (5\text{-}5a)$$

$$N_{qp} = \frac{3}{3 - \sin\phi} \cdot e^{\frac{(90-\phi')\pi}{180}\tan\phi'}$$

$$\cdot \tan^2\left[45 + \frac{\phi'}{2}\right] \cdot I_{rr}^{\frac{4\sin\phi'}{3(1+\sin\phi')}} \qquad (5\text{-}5b)$$

$$I_{rr} = \frac{I_r}{1 + \epsilon_v \cdot I_r} \qquad (5\text{-}5c)$$

$$I_r = \frac{G_s}{c + \sigma_L' \tan\phi'} \qquad (5\text{-}5d)$$

$$\epsilon_v = \frac{1 - 2\upsilon_s}{2(1 - \upsilon_s)} \cdot \frac{\sigma_L'}{G_s} \qquad (5\text{-}5e)$$

where

I_{rr} = reduced rigidity index

I_r = rigidity index

ϵ_v = volumetric strain, fraction

υ_s = soil Poisson's ratio

G_s = soil shear modulus, ksf

c = undrained shear strength C_u, ksf

ϕ' = effective friction angle, deg

σ_L' = effective soil vertical overburden pressure at pile base, ksf

$I_{rr} \approx I_r$ for undrained or dense soil where $\upsilon_s \approx 0.5$. G_s may be estimated from laboratory or field test data, Chapter 3, or by methods described in EM 1110-1-1904. The shape factor $\zeta_{cp} = 1.00$ and

$$\zeta_{qp} = \frac{1 + 2K_o}{3} \qquad (5\text{-}6a)$$

$$K_o = (1 - \sin\phi') \cdot OCR^{\sin\phi'} \qquad (5\text{-}6b)$$

where

K_o = coefficient of earth pressure at rest

OCR = overconsolidation ratio

The OCR can be estimated by methods described in Chapter 3 or EM 1110-1-1904. If the OCR is not known, the Jaky equation can be used

$$K_o = 1 - \sin\phi' \qquad (5\text{-}6c)$$

c. Vesic Alternate Method. A conservative estimate of N_{qp} can be readily made by knowing only the value of ϕ'

$$N_{qp} = (1 + \tan\phi') \cdot e^{\tan\phi'} \cdot \tan^2\left[45 + \frac{\phi'}{2}\right] \qquad (5\text{-}7)$$

The shape factor ζ_{qp} may be estimated by Equation 5-6. Equation 5-7 assumes a local shear failure and hence leads to a lower bound estimate of q_{bu}. A local shear failure can occur in poor soils such as loose silty sands or weak clays or else in soils subject to disturbance.

d. General Shear Method. The bearing capacity factors of Equation 5-2b may be estimated assuming general shear failure by (Bowles 1968)

$$N_{qp} = \frac{e^{\frac{270-\phi'}{180}\pi\tan\phi'}}{2 \cdot \cos^2\left[45 + \frac{\phi'}{2}\right]} \qquad (5\text{-}8)$$

The shape factor $\zeta_{qp} = 1.00$. $N_{cp} = (N_{qp} - 1)\cot \phi'$.

B. SKIN FRICTION CAPACITY. The maximum skin friction that may be mobilized along an el-

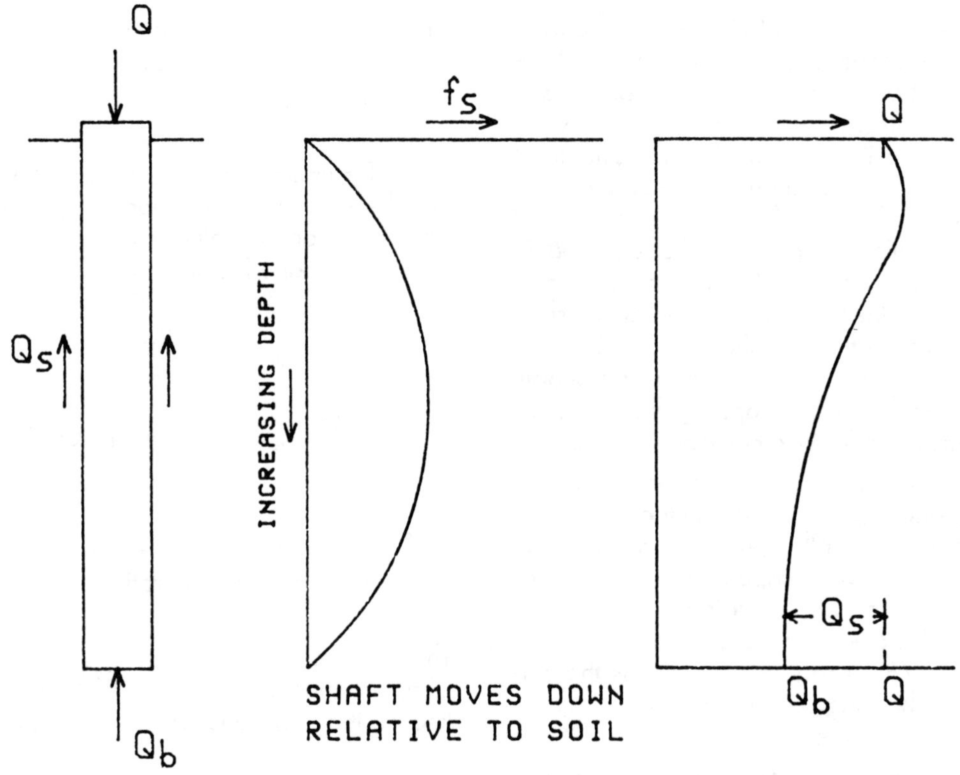

Figure 5-4. An example distribution of skin friction in a pile

ement of shaft length ΔL may be estimated by

$$Q_{sui} = A_{si} \cdot f_{si} \qquad (5\text{-}9)$$

where

A_{si} = surface area of element i, $C_{si} \cdot \Delta L$, ft^2

C_{si} = shaft circumference at element i, ft

ΔL = length of pile element, ft

f_{si} = skin friction at pile element i, ksf

Resistance to applied loads from skin friction along the shaft perimeter increases with increasing depth to a maximum, then decreases toward the tip. One possible distribution of skin friction is indicated in Figure 5-4. The estimates of skin friction f_{si} with depth is at best approximate. Several methods of estimating f_{si}, based on past experience and the results of load tests, are described below. The vertical load on the shaft may initially increase slightly with increasing depth near the ground surface because the pile adds weight which may not be supported by the small skin friction near the surface. Several of these methods should be used when possible to provide a range of probable skin friction values.

1. Cohesive Soil. Adhesion of cohesive soil to the shaft perimeter and the friction resisting applied loads are influenced by the soil shear strength, soil disturbance and changes in pore pressure, and lateral earth pressure existing after installation of the deep foundation. The average undrained shear strength determined from the methods described in Chapter 3 should be used to estimate skin friction. The friction angle ϕ is usually taken as zero.

a. The soil-shaft skin friction f_{si} of a length of shaft element may be estimated by

$$f_{si} = \alpha_a \cdot C_u \qquad (5\text{-}10)$$

where

α_a = adhesion factor

C_u = undrained shear strength, ksf

Local experience with existing soils and load test results should be used to estimate appropriate α_a. Estimates of α_a may be made from Table 5-1 in the absence of load test data and for preliminary design.

b. The adhesion factor may also be related to the plasticity index PI for drilled shafts constructed dry by (Data from Stewart and Kulhawy 1981)

Overconsolidated: $\alpha_a = 0.7 - 0.01 \cdot PI \qquad (5\text{-}11a)$

Table 5-1. Adhesion Factors for Drilled Shafts in a Cohesive Soil (Reese and O'Neill 1988)

Shaft Depth, ft	Adhesion Factor α_a
0 – 5	0.0
Shaft diameter from bottom of straight shaft or from top of underream	0.0
All Other Points	0.55

Note: skin friction f_{si} should be limited to 5.5 ksf

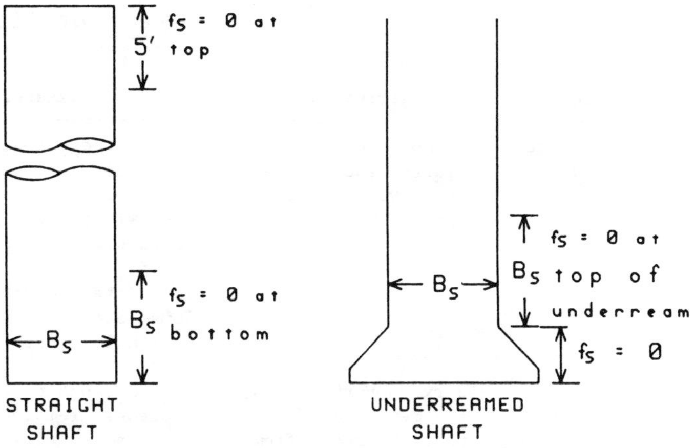

Slightly over-consolidated (OCR ≤ 2): $\alpha_a = 0.9 - 0.01 \cdot PI$ (5-11b)

Normally consolidated: $\alpha_a = 0.9 - 0.004 \cdot PI$ (5-11c)

where $15 < PI < 80$. Drilled shafts constructed using the bentonite-water slurry should use α_a of about 1/2 to 2/3 of those given by Equations 5-11.

2. Cohesionless Soil. The soil-shaft skin friction may be estimated using effective stresses with the beta method

$$f_{si} = \beta_f \cdot \sigma_i' \quad (5\text{-}12a)$$

$$\beta_f = K \cdot \tan \delta_a \quad (5\text{-}12b)$$

where

β_f = lateral earth pressure and friction angle factor

K = lateral earth pressure coefficient

δ_a = soil-shaft effective friction angle, ≤ ϕ', degrees

σ_i' = effective vertical stress in soil in shaft (pile) element i, ksf

The cohesion c is taken as zero.

a. Figure 5-5 indicates appropriate values of β_f as a function of the effective friction angle ϕ' of the soil prior to installation of the deep foundation.

b. Refer to Figure 5-3 to determine the critical depth Lc below which σ_i' remains constant with increasing depth.

3. CPT Field Estimate. The skin friction f_{si} may be estimated from the measured cone resistance q_c for the piles described in Table 5-2 using the curves given in Figure 5-6 for clays and silt, sands and gravels, and chalk (Bustamante and Gianeselli 1983).

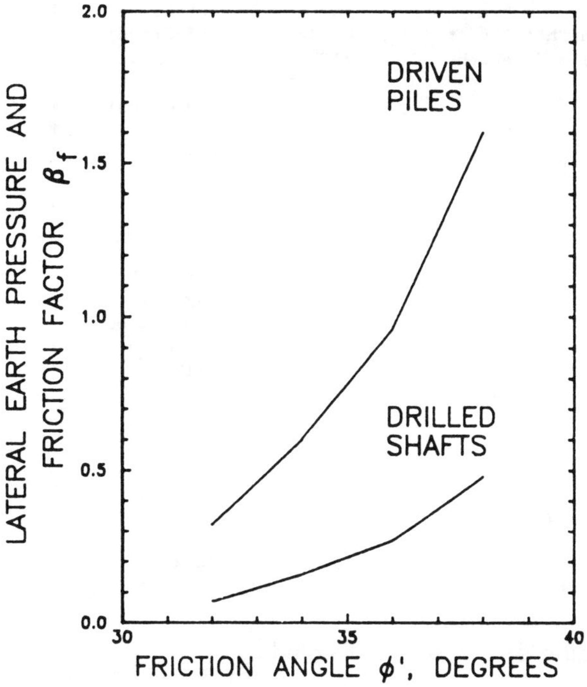

Figure 5-5. Lateral earth pressure and friction angle factor β as a function of friction angle prior to installation (Data from Meyerhof 1976 and Poulos and Davis 1980)

C. EXAMPLE APPLICATION. A 1.5-ft diameter straight concrete drilled shaft is to be constructed 30 ft deep through a 2-layer soil of a slightly overconsolidated clay with $PI = 40$ and fine uniform sand (Figure 5-7). Depth of embedment in the sand layer $L_b = 15$ ft. The water table is 15 ft below ground

Table 5-2. Descriptions of Deep Foundations. Note that the curves matching the numbers are found in Figure 5-6. (Data from Bustamante and Gianeselli 1983)

a. Drilled Shafts

Pile	Description	Remarks	Cone Resistance q_c, ksf	Soil	Curve
Drilled shaft bored dry	Hole bored dry without slurry; applicable to cohesive soil above water table	Tool without teeth; oversize blades; remolded soil on sides	any	Clay-Silt	1
		Tool with teeth; immediate concrete placement	> 25	Clay-Silt	2
			> 94	Clay-Silt	3
			any	Chalk	1
		Immediate concrete placement	> 94	Chalk	3
		Immediate concrete placement with load test	>250	Chalk	4
Drilled shaft with slurry	Slurry supports sides; concrete placed through tremie from bottom up displacing	Tool without teeth; oversize blades; remolded soil on sides	any	Clay-Silt	1
		Tool with teeth; immediate concrete placement	> 25	Clay-Silt	2
			> 94	Clay-Silt	3
			any	Sand-Gravel	1
		Fine sands and length < 100 ft	>104	Sand-Gravel	2
		Coarse gravelly sand/gravel and length < 100 ft	>156	Sand-Gravel	3
		Gravel	> 83	Sand-Gravel	4
			any	Chalk	1
		Above water table; immediate concrete placement	> 94	Chalk	3
		Above water table; immediate concrete placement with load test	>250	Chalk	4
Drilled shaft with casing	Bored within steel casing; concrete placed as casing retrieved		any	Clay-Silt	1
		Dry holes	> 25	Clay-Silt	2
			any	Sand-Gravel	1
		Fine sands and length < 100 ft	> 104	Sand-Gravel	2
		Coarse sand/gravel and length < 100 ft	>157	Sand-Gravel	3
		Gravel	> 83	Sand-Gravel	4
			any	Chalk	1
		Above water table; immediate concrete placement	> 94	Chalk	3
		Above water table; immediate concrete placement	>250	Chalk	4
Drilled shaft hollow auger (auger cast pile)	Hollow stem continuous auger length > shaft length; auger extracted without turning while concrete injected through auger stem		any	Clay-Silt	1
			> 25	Clay-Silt	2
			any	Sand-Gravel	1
		Sand exhibiting some cohesion	>104	Sand-Gravel	2
			any	Chalk	1
Pier	Hand excavated; sides supported with retaining elements or casing		any	Clay-Silt	1
			> 25	Clay-Silt	2
		Above water table; immediate concrete placement	> 94	Chalk	3
		Above water table; immediate concrete placement	>250	Chalk	4

surface at the clay-sand interface. The concrete unit weight γ_{conc} = 150 lbs/ft^3. Design load Q_d = 75 kips.

1. Soil Parameters.

a. The mean effective vertical stress in a soil layer σ'_s such as in a sand layer below a surface layer, Figure 5-7, may be estimated by

$$\sigma'_s = L_{clay} \cdot \gamma'_c + \frac{L_{sand}}{2} \cdot \gamma'_s \qquad (5\text{-}13a)$$

Table 5-2. (Continued)

Pile	Description	Remarks	Cone Resistance q_c, ksf	Soil	Curve
Micropile I	Drilled with casing; diameter < 10 in.; casing recovered by applying pressure inside top of plugged casing		any	Clay-Silt	1
			> 25	Clay-Silt	2
		With load test	> 25	Clay-Silt	3
			any	Sand-Gravel	1
		Fine sands with load test	>104	Sand-Gravel	2
		Coarse gravelly sand/gravel	>157	Sand-Gravel	3
			any	Chalk	1
			> 94	Chalk	3
Micropile II	Drilled < 10 in. diameter; reinforcing cage placed in hole and concrete placed from bottom-up		any	Clay-Silt	1
			> 42	Clay-Silt	4
		With load test	> 42	Clay-Silt	5
			>104	Sand-Gravel	5
			> 94	Chalk	4
High pressure injected	Diameter > 10 in. with injection system capable of high pressures		any	Clay-Silt	1
			> 42	Clay-Silt	5
			>104	Sand-Gravel	5
		Coarse gravelly sand/gravel	>157	Sand-Gravel	3
			> 94	Chalk	4

b. Driven Piles

Pile	Description	Remarks	Cone Resistance q_c, ksf	Soil	Curve
Screwed-in	Screw type tool placed in front of corrugated pipe that is pushed or screwed in place; reverse rotation to pull casing while placing concrete		any	Clay-Silt	1
		q_c < 53 ksf	> 25	Clay-Silt	2
		Slow penetration	> 94	Clay-Silt	3
		Slow penetration	any	Sand-Gravel	1
		Fine sands with load test	> 73	Sand-Gravel	2
		Coarse gravelly sand/gravel	>157	Sand-Gravel	3
		Coarse gravelly sand/gravel	any	Chalk	1
		q_c < 146 ksf without load test	> 63	Chalk	2
		q_c < 146 ksf with load test	> 63	Chalk	3
		Above water table; immediate concrete placement; slow penetration	> 94	Chalk	3
		Above water table with load test	>250	Chalk	4
Concrete coated	6 to 20 in. diameter pipe; H piles; caissons of 2 to 4 sheet pile sections; pile driven with oversize protecting shoe; concrete injected through hose near oversize shoe producing coating around pile		any	Clay-Silt	1
			any	Sand-Gravel	1
			>157	Sand-Gravel	4
			any	Chalk	1
		With load test	> 63	Chalk	3
			> 94	Chalk	3
			>250	Chalk	4

Table 5-2. (Continued)

Pile	Description	Remarks	Cone Resistance q_c, ksf	Soil	Curve
Prefabricated	Reinforced or		any	Clay-Silt	1
	prestressed concrete		any	Sand-Gravel	1
	installed by driving	Fine Sands	>157	Sand-Gravel	2
	or vibrodriving	Coarse gravelly sand/gravel	>157	Sand-Gravel	3
		With load test	>157	Sand-Gravel	4
			any	Chalk	1
		q_c < 147 ksf without load test	> 63	Chalk	2
		q_c < 147 ksf with load test	> 63	Chalk	3
		With load test	>250	Chalk	4
Steel	H piles; pipe piles;		any	Clay-Silt	1
	any shape obtained by		any	Sand-Gravel	1
	welding sheet-pile	Fine sands with load test	> 73	Sand-Gravel	2
	sections	Coarse gravelly sand/gravel	>157	Sand-Gravel	3
			any	Chalk	1
		q_c < 147 ksf without load test	> 63	Chalk	2
		q_c < 147 ksf with load test	> 63	Chalk	3
Prestressed tube	Hollow cylinder element		any	Clay-Silt	1
	of lightly reinforced		any	Sand-Gravel	1
	concrete assembled by	With load test	> 73	Sand-Gravel	2
	prestressing before	Fine sands with load test	>157	Sand-Gravel	2
	driving; 4-9 ft long	Coarse gravelly sand/gravel	>157	Sand-Gravel	3
	elements; 2-3 ft	With load test	>157	Sand-GRavel	4
	diameter; 6 in.		< 63	Chalk	1
	thick; piles driven	q_c < 146 ksf	> 63	Chalk	2
	open ended				
		With load test	> 63	Chalk	3
		With load test	>250	Chalk	4
Concrete plug bottom of Pipe	Driving accomplished		any	Clay-Silt	1
	through bottom	q_c < 42 ksf	> 25	Clay-Silt	3
	concrete plug;		any	Sand-Gravel	1
	casing pulled	Fine sands with load test	> 73	Sand-Gravel	2
	while low slump		any	Chalk	1
	concrete compacted		> 94	Chalk	4
	through casing				
Molded	Plugged tube driven to		any	Clay-Silt	1
	final position; tube	With load test	> 25	Clay-Silt	2
	filled to top with		any	Sand-Gravel	1
	medium slump concrete	Fine sand with load test	> 73	Sand-Gravel	2
	and tube extracted	Coarse gravelly sand/gravel	>157	Sand-Gravel	3
			any	Chalk	1
		q_c < 157 ksf	> 63	Chalk	2
		With load test	> 63	Chalk	3
		With load test	>250	Chalk	4
Pushed-in concrete	Cylindrical concrete		any	Clay-Silt	1
	elements prefabricated		any	Sand-Gravel	1
	or cast-in-place	Fine sands	>157	Sand-Gravel	2
	1.5-8 ft long, 1-2 ft	Coarse gravelly sand/gravel	>157	Sand-Gravel	3
	diameter; elements	Coarse gravelly sand/gravel with			
	pushed by hydraulic	load test	>157	Sand-Gravel	4
	jack		any	Chalk	1
		q_c < 157 ksf	> 63	Chalk	2
		With load test	> 63	Chalk	3
		With load test	>250	Chalk	4

Table 5-2. (Concluded)

Pile	Description	Remarks	Cone Resistance q_c, ksf	Soil	Curve
Pushed-in steel	Steel piles pushed in by hydraulic jack		any	Clay-Silt	1
			any	Sand-Gravel	1
		Coarse gravelly sand/gravel	>157	Sand-Gravel	3
			any	Chalk	1
		q_c < 157 ksf	> 63	Chalk	2
		With load test	>250	Chalk	4

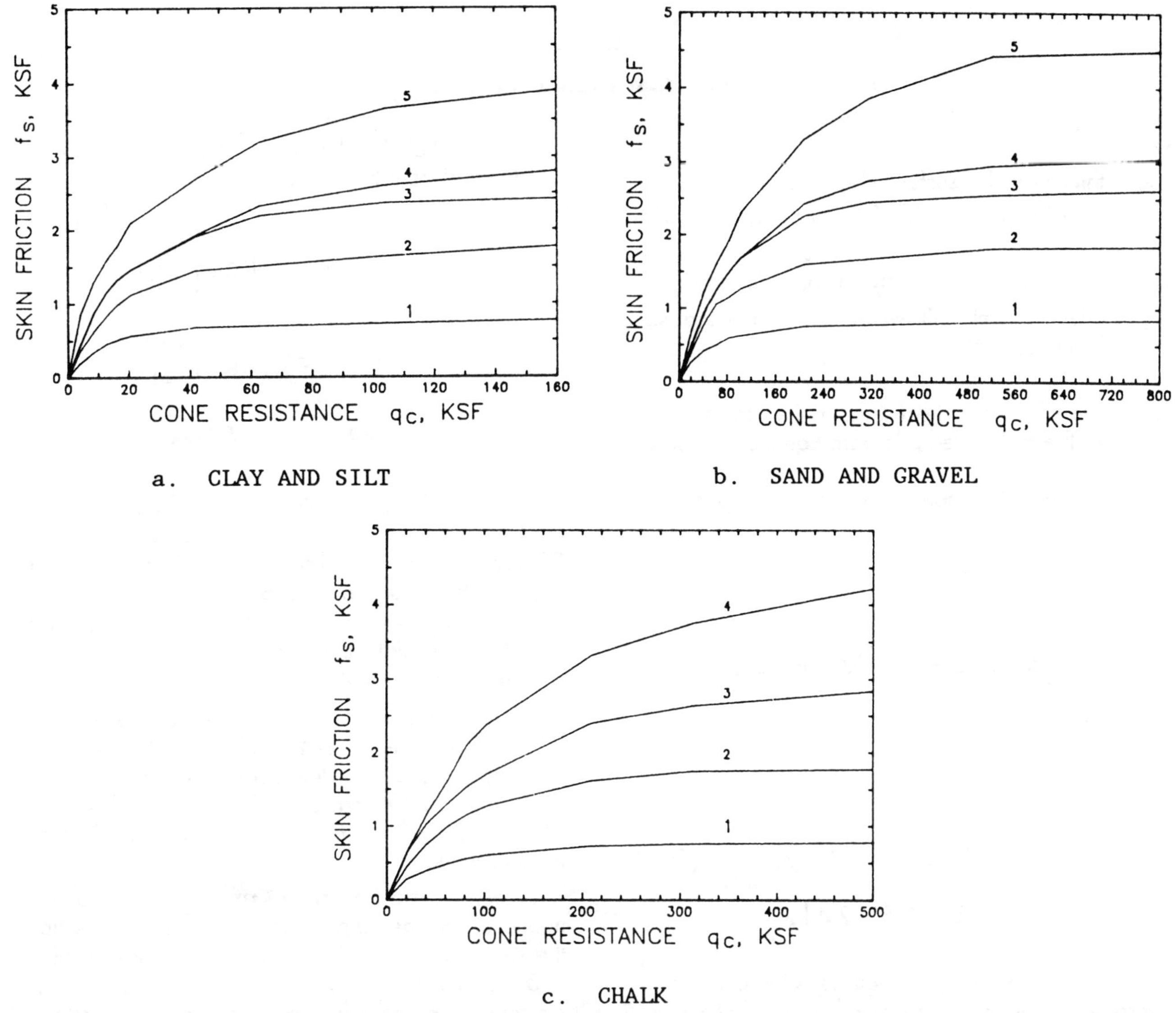

Figure 5-6. Skin friction and cone resistance relationships for deep foundations (Data from Bustamante and Gianeselli 1983). The appropriate curve to use is determined from Table 5-2

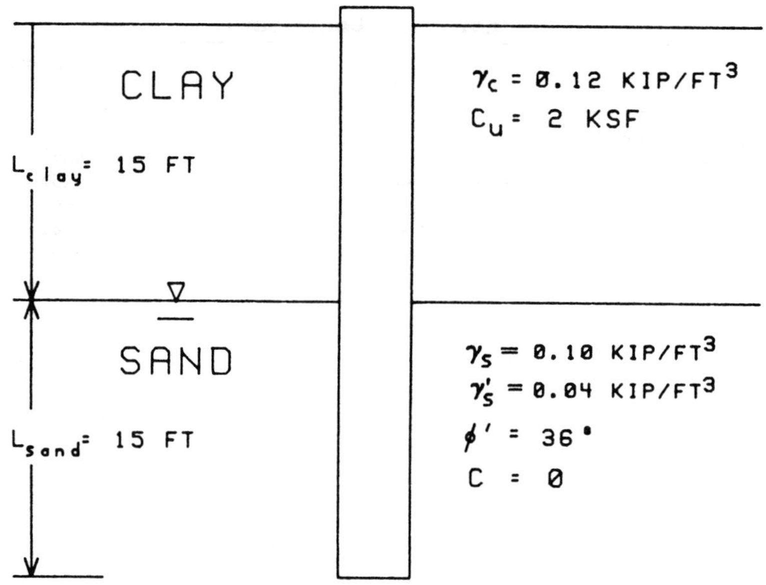

Figure 5-7. Drilled shaft 1.5-ft diameter at 30-ft depth

where

L_{clay} = thickness of a surface clay layer, ft

γ_c' = effective unit weight of surface clay layer, kips/ft^3

L_{sand} = thickness of an underlying sand clay layer, ft

γ_s' = effective wet unit weight of underlying sand layer, kips/ft^3

The mean effective vertical stress in the sand layer adjacent to the embedded pile from Equation 5-13a is

The effective vertical soil stress at the pile tip is

$$\sigma_s' = L_{clay} \cdot \gamma_c' + \frac{L_{sand}}{2} \cdot \gamma_s'$$

$$= 15 \cdot 0.12 + \frac{15}{2} \cdot 0.04 = 1.8 + 0.3$$

$$\sigma_s' = 2.1 \text{ ksf}$$

$$\sigma_L' = L_{clay} \cdot \gamma_c' + L_{sand} \cdot \gamma_s'$$

$$= 15 \cdot 0.12 + 15 \cdot 0.04 \qquad (5\text{-}13b)$$

$$= 1.8 + 0.6 = 2.4 \text{ ksf}$$

b. Laboratory strength tests indicate that the average undrained shear strength of the clay is $C_u = 2$ ksf. Cone penetration tests indicate an average cone tip resistance q_c in the clay is 40 ksf and in the sand 160 ksf.

c. Relative density of the sand at the shaft tip is estimated from Equation 3-5

$$D_r = -74 + 66 \cdot \log_{10} \frac{q_c}{(\sigma_{vo}')^{0.5}}$$

$$= -74 + 66 \cdot \log_{10} \frac{160}{(2.4)^{0.5}}$$

$$D_r = -74 + 133 = 59 \text{ percent}$$

The effective friction angle estimated from Table 3-1a is $\phi' = 38$ deg, while Table 3-1b indicates $\phi' = 36$ to 38 deg. Figure 3-1 indicates $\phi' = 38$ deg. The sand appears to be of medium to dense density. Select a conservative $\phi = 36$ deg. Coefficient of earth pressure at rest from the Jaky Equation 5-6c is $K_o = 1 - \sin\phi = 1 - \sin 36 \text{ deg} = 0.42$.

d. The sand elastic modulus E_s is at least 250 ksf from Table D-3 in EM 1110-1-1904 using guidelines for a medium to dense sand. The shear modulus G_s is estimated using $G_s = E_s/[2(1 + \upsilon_s)] = 250/[2(1 + 0.3)] = 96$ or approximately 100 ksf. Poisson's ratio of the sand $\upsilon_s = 0.3$.

2. End Bearing Capacity. A suitable estimate of end bearing capacity q_{bu} for the pile tip in the sand may be evaluated from the various methods in 5-2a for cohesionless soil as described below. Hansin and Vesic methods account for a limiting effective stress, while the general shear method and Vesic alternate method ignore this stress. The Vesic Alternate method is not used because the sand appears to be of medium density and not loose. Local shear failure is not likely.

a. Hansen Method. From Table 4-4 (or calculated from Table 4-5), $N_{\gamma p} = 37.75$ and $N_p = 40.05$ for $\phi' = 36$ deg. From Table 4-5,

$$\zeta_{qs} = 1 + \tan\phi = 1 + \tan 36 = 1.727$$

$$\zeta_{qd} = 1 + 2\tan\phi\,(1 - \sin\phi)^2 \cdot \tan^{-1}(L_{sand}/B)$$

$$= 1 + 2\tan 36\,(1 - \sin 36)^2 \cdot \tan^{-1}(15/1.5) \cdot \pi/180$$

$$= 1 + 2 \cdot 0.727(1 - 0.588)^2 \cdot 1.471 = 1.363$$

$$\zeta_{qp} = \zeta_{qs} \cdot \zeta_{qd} = 1.727 \cdot 1.363 = 2.354$$

$$\zeta_{\gamma s} = 1 - 0.4 = 0.6$$

$$\zeta_{\gamma d} = 1.00$$

$$\zeta_{\gamma p} = \zeta_{\gamma s} \cdot \zeta_{\gamma d} = 0.6 \cdot 1.00 = 0.6$$

From Equation 5-2a

$$q_{bu} = \sigma'_L \cdot N_{qp}\zeta_{qp} + \frac{B_b}{2} \cdot \gamma'_b \cdot N_{\gamma p} \cdot \zeta_{\gamma p}$$

$$= 2.4 \cdot 37.75 \cdot 2.354 + \frac{1.5}{2}$$

$$\cdot\, 0.04 \cdot 40.05 \cdot 0.6$$

$$= 213.3 + 0.7 = 214 \text{ ksf}$$

b. Vesic Method. The reduced rigidity index from Equation 5-5c is

$$I_{rr} = \frac{I_r}{1 + \epsilon_v \cdot I_r} = \frac{57.3}{1 + 0.006 \cdot 57.3} = 42.6$$

where

$$\epsilon_v = \frac{1 - 2\upsilon_s}{2(1 - \upsilon_s)} \cdot \frac{\sigma'_L}{G_s}$$

$$= \frac{1 - 2 \cdot 0.3}{2(1 - 0.3)} \cdot \frac{2.4}{100} = 0.006 \qquad \text{(5-5e)}$$

$$I_r = \frac{G_s}{c + \sigma'_L \tan\phi'}$$

$$= \frac{100}{2.4 \cdot \tan 36} = 57.3 \qquad \text{(5-5d)}$$

From Equation 5-5b

$$N_{qp} = \frac{3}{3 - \sin\phi} \cdot e^{\frac{(90-\phi')\pi}{180}\tan\phi'}$$

$$\cdot \tan^2\left[45 + \frac{\phi'}{2}\right] \cdot I_{rr}^{\frac{4\sin\phi'}{3(1-\sin\phi')}}$$

$$= \frac{3}{3 - \sin 36} \cdot e^{\frac{(90-36)\pi}{180}\tan 36}$$

$$\cdot \tan^2\left[45 + \frac{36}{2}\right] \cdot I_{rr}^{\frac{4\sin 36}{3(1+\sin 36)}}$$

$$N_{qp} = \frac{3}{3 - 0.588} \cdot e^{0.685} \cdot 3.852 \cdot 42.6^{0.494}$$

$$= 1.244 \cdot 1.984 \cdot 3.852 \cdot 6.382 = 60.7$$

The shape factor from Equation 5-6a is

$$\zeta_{qp} = \frac{1 + 2K_o}{3} = \frac{1 + 2 \cdot 0.42}{3} = 0.61$$

From Equation 5-2c,

$$q_{bu} = \sigma'_L \cdot N_{qp} \cdot \zeta_{qp} = 2.4 \cdot 60.7 \cdot 0.61 = 88.9 \text{ ksf}$$

c. General Shear Method. From Equation 5-8

$$N_{qp} = \frac{e^{\frac{270-\phi'}{180}\pi\tan\phi'}}{2 \cdot \cos^2\left[45 + \frac{\phi'}{2}\right]} = \frac{e^{\frac{270-36}{180}\kappa \cdot \tan 36}}{2 \cdot \cos^2\left[45 + \frac{36}{2}\right]}$$

$$= \frac{e^{1.3\pi \cdot 0.727}}{2 \cdot 0.206} = 47.24$$

The shape factor $\zeta_{qp} = 1.00$ when using Equation 5-8. From Equation 5-2c,

$$q_{bu} = \sigma'_L \cdot N_{qp} \cdot \zeta_{qp} = 2.4 \cdot 47.24 \cdot 1.00 = 113.4 \text{ ksf}$$

d. Comparison of Methods. A comparison of methods is shown as follows:

Method	q_{bu}, ksf
Hansen	214
Vesic	89
General Shear	113

The Hansen result of 214 ksf is much higher than the other methods and should be discarded without proof from a load test. The Vesic and General Shear methods give an average value q_{bu} = 102 ksf.

2. Skin Friction Capacity. A suitable estimate of skin friction f_s from the soil-shaft interface may be evaluated by methods in Section 5-2b for embedment of the shaft in both clay and sand as illustrated below.

a. Cohesive Soil. The average skin friction from Equation 5-10 is

$$f_s = \alpha_a \cdot C_u = 0.5 \cdot 2 = 1.0 \text{ ksf}$$

where α_a was estimated from Equation 5-11b, $\alpha_a = 0.9 - 0.01 \cdot 40 = 0.5$ or 0.55 from Table 5-1. Skin friction from the top 5 ft should be neglected.

b. Cohesionless Soil. Effective stresses are limited by $Lc/B = 10$ or to depth $Lc = 15$ ft. Therefore, $\sigma'_s = 1.8$ ksf, the effective stress at 15 ft. The average skin friction from Equation 5-12a is

$$f_s = \beta_f \cdot \sigma'_s = 0.26 \cdot 1.8 = 0.5 \text{ ksf}$$

where $\beta_f = 0.26$ from Figure 5-5 using $\phi' = 36$ deg.

c. CPT Field Estimate. The shaft was bored using bentonite-water slurry. Use curve 2 of clay and silt, Figure 5-6a, and curve 3 of sand and gravel, Figure 5-6b. From these figures, f_s of the clay is 1.5 ksf and f_s of the sand is 2.0 ksf.

d. Comparison of Methods. Skin friction varies from 1.0 to 1.5 ksf for the clay and 0.5 to 2 ksf for the sand. Skin friction is taken as 1 ksf in the clay and 1 ksf in the sand, which is considered conservative.

3. Total Capacity. The total bearing capacity from Equation 5-1a is

$$Q_u = Q_{bu} + Q_{su} - W_p$$

where

$$W_p = \frac{\pi B^2}{4} \cdot L_{clay} \cdot \gamma_{conc} + \frac{\pi B^2}{4} \cdot L_{sand} \cdot \gamma'_{conc}$$

$$= \frac{\pi 1.5^2}{4} \cdot \left[15 \cdot \frac{50}{1000} + 15 \cdot \frac{86}{1000}\right]$$

$$= 4.0 + 2.3 = 6.3 \text{ kips}$$

γ_{conc} is the unit weight of concrete, 150 lbs/ft^3.

a. Q_{bu} from Equation 5-1b is

$$Q_{bu} = q_{bu} \cdot A_b = 102 \cdot 1.77 = 180 \text{ kips}$$

where A_b = area of the base, $\pi B^2/4 = \pi \cdot 1.5^2/4 = 1.77$ ft^2.

b. Q_{su} from Equation 5-1b and 5-9

$$Q_{su} = \sum_{i=1}^{n} Q_{sui} = C_s \cdot \Delta L \sum_{i=1}^{2} f_{si}$$

where $C_s = \pi B$ and $\Delta L = 15$ ft for clay and 15 ft for sand. Therefore,

$$Q_{su} = \pi \cdot \underset{\text{sand}}{B} \cdot [\Delta L \cdot \underset{\text{clay}}{f_s} + \Delta L \cdot f_s)$$

$$= \pi \cdot 1.5 \cdot [15 \cdot 1 + 10 \cdot 1)] = 118 \text{ kips}$$

where skin friction is ignored in the top 5 ft of clay.

c. Total Capacity. Inserting the end bearing and skin resistance bearing capacity values into Equation 5-1a is

$$Q_u = 180 + 118 - 6 = 292 \text{ kips}$$

4. Allowable Bearing Capacity. The allowable bearing capacity from Equation 1-2b is

$$Q_a = \frac{Q_u}{FS} = \frac{292}{3} = 97 \text{ kips}$$

using $FS = 3$ from Table 1-2. $Q_d = 75 < Q_a = 97$ kips. A settlement analysis should also be performed to check that settlement is tolerable. A load test is recommended to confirm or correct the ultimate bearing capacity. Load tests can increase Q_a because $FS = 2$ and permit larger Q_d depending on results of settlement analysis.

D. LOAD TESTS FOR VERTICAL CAPACITY. ASTM D 1143 testing procedures for piles under static compression loads are recommended and should be performed for individual or groups of vertical and batter shafts (or piles) to determine their response to axially-applied compression loads. Load tests lead to the most efficient use of shafts or piles. The purpose of testing is to verify that the actual pile response to compression loads corresponds to that used for design and that the calculated ultimate load is less than the actual ultimate load. A load cell placed at the bottom of the shaft can be used to determine the end bearing resis-

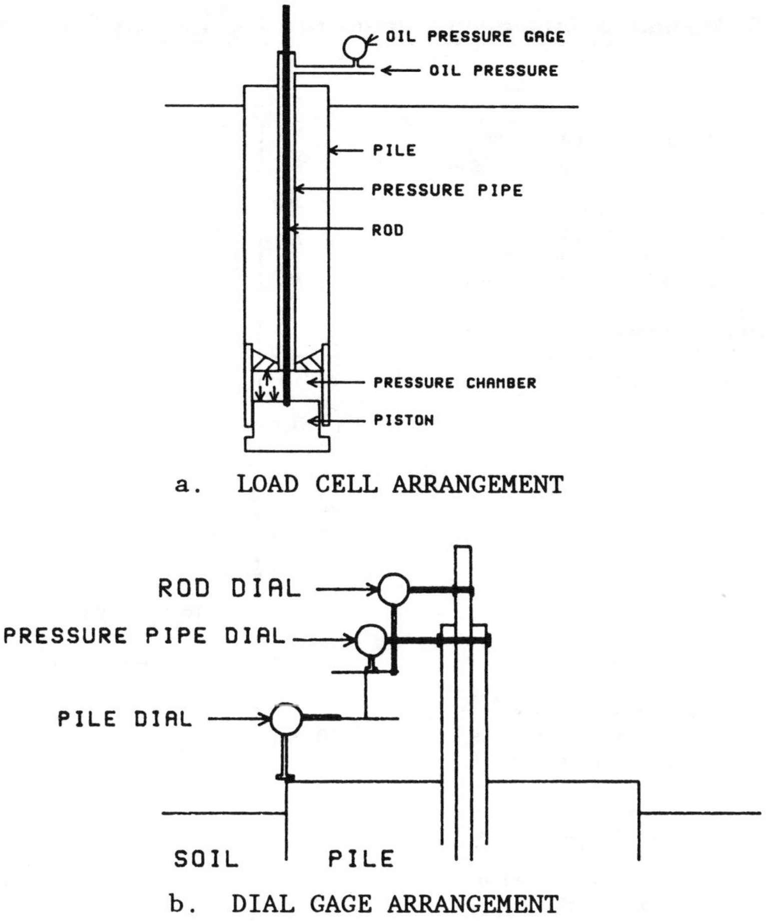

Figure 5-8. Example load test arrangement for Osterberg method

tance and to calculate skin friction as the difference between the total capacity and end bearing resistance.

1. Quick Load Test. The Quick Load Test option is normally satisfactory, except that this test should be taken to plunging failure or three times the design load or 1,000 tons, whichever comes first.

2. Cost Savings. Load tests can potentially lead to significant savings in foundation costs, particularly on large construction projects when a substantial part of the bearing capacity is contributed by skin friction. Load tests also assist the selection of the best type of shaft or pile and installation depth.

3. Lower Factor of Safety. Load tests allow use of a lower safety factor of 2 and can offer a higher allowable capacity.

4. Scheduling of Load Tests. Load tests are recommended during the design phase, when economically feasible, to assist in selection of optimum equipment for construction and driving the piles in addition to verifying the bearing capacity. This information can reduce contingency costs in bids and reduce the potential for later claims.

a. Load tests are recommended for most projects during early construction to verify that the allowable loads used for design are appropriate and that installation procedures are satisfactory.

b. Load tests during the design phase are economically feasible for large projects such as for multistory structures, power plants, locks and dams.

c. When load tests are performed during the design phase, care must be taken to ensure that the same procedures and equipment (or driving equipment including hammer, helmet, cushion, etc. in the case of driven piles) are used in actual construction.

5. Alternative Testing Device. A load testing device referred to as the Osterberg method (Osterberg 1984) can be used to test both driven piles and drilled shafts. A piston is placed at the bottom of the bored shaft before the concrete is placed or the piston can be mounted at the bottom of a pile (Figure 5-8a).

Table 5-3. Methods of Estimating Ultimate Bearing Capacity From Load Tests

Method	Procedure	Diagram
Tangent (Butler and Hoy 1977)	1. Draw a tangent line to the curve at the graph's origin 2. Draw another tangent line to the curve with slope equivalent to slope of 1 inch for 40 kips of load 3. Ultimate bearing capacity is the load at the intersection of the tangent lines	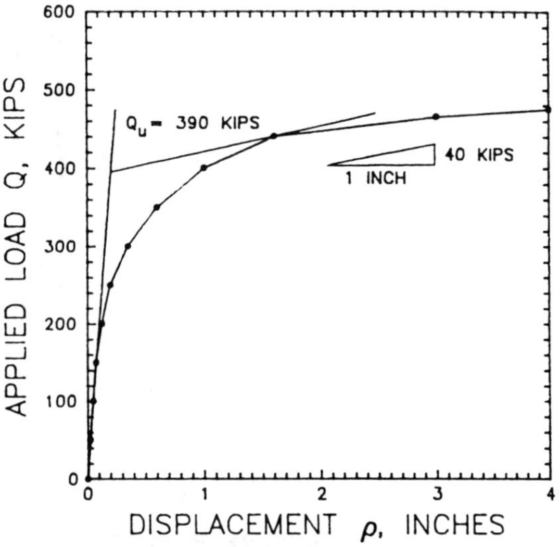
Limit Value (Davisson 1972)	1. Draw a line with slope $\frac{\pi B^2}{4L} \cdot E_p$ where B = diameter of pile, inches; E_p = Young's pile modulus, kips/inch^2; L = pile length, inches 2. Draw a line parallel with the first line starting at a displacement of 0.15 + B/120 inch at zero load 3. Ultimate bearing capacity is the load at the intersection of the load-displacement curve with the line of step 2	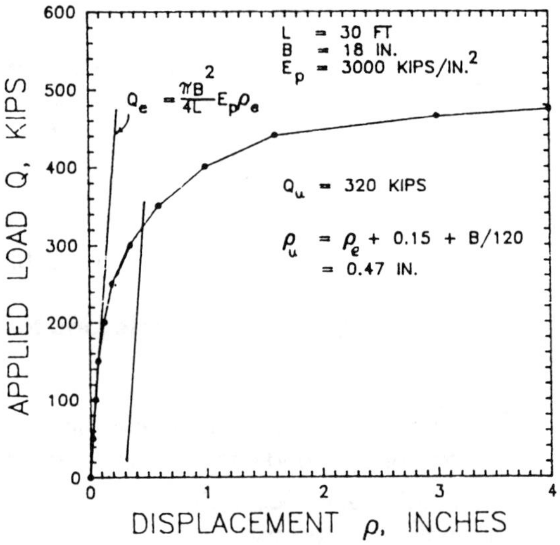

Pressure is applied to hydraulic fluid which fills a pipe leading to the piston. Fluid passes through the annular space between the rod and pressure pipe into the pressure chamber. Hydraulic pressure expands the pressure chamber forcing the piston down. This pressure is measured by the oil (fluid) pressure gage, which can be calibrated to determine the force applied to the bottom of the pile and top of the piston. End bearing capacity can be determined if the skin friction capacity exceeds the end bearing capacity. This condition is frequently not satisfied.

a. A dial attached to the rod with the stem on the reference beam (Figure 5-8b) measures the downward movement of the piston. A dial attached to the pressure pipe measures the upward movement of the pile base. A third dial attached to the reference beam with stem on the pile top measures the movement of the pile top. The difference in readings between the top and bottom of the pile is the elastic compression due to side friction. The total side friction force can be estimated using Young's modulus of the pile.

b. If the pile is tested to failure, the measured force at failure (piston downward movement is continuous with time or excessive according to guidance in Table 5-3) is the ultimate end bearing capacity. The measured failure force in the downward plunging piston therefore provides a *FS* > 2 against failure considering that the skin friction capacity is equal to or greater than the end bearing capacity.

c. This test can be more economical and completed more quickly than a conventional load test; friction and end bearing resistance can be determined

Table 5-3. (Concluded)

<table>
<tr><th>Method</th><th>Procedure</th><th>Diagram</th></tr>
<tr><td>80 Percent (Hansen 1963)</td><td>1. Plot load test results as $\frac{\sqrt{\rho}}{Q}$ vs. ρ
2. Draw straight line through data points
3. Determine the slope a and intercept b of this line
4. Ultimate bearing capacity is
$Q_u = \frac{1}{2\sqrt{ab}}$
5. Ultimate deflection is
$\rho_u = b/a$</td><td>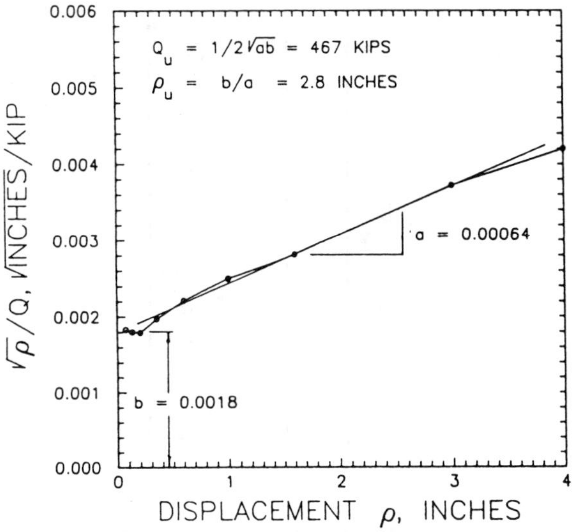
</td></tr>
<tr><td>90 Percent (Hansen 1963)</td><td>1. Calculate 0.9Q for each load Q
2. Find $\rho_{0.9Q}$, displacement for load of 0.9Q, for each Q from Q vs. ρ plot
3. Determine $2\rho_{0.9Q}$ for each Q and plot vs. Q on chart with the load test data of Q vs. ρ
4. Ultimate bearing capacity is the load at the intersection of the two plots of data</td><td>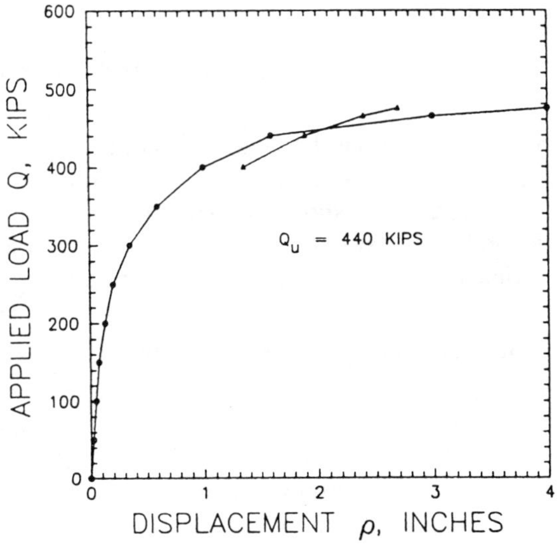
</td></tr>
</table>

separately; optimum length of driven piles can be determined by testing the same pile at successfully deeper depths. Other advantages include ability to work over water, to work in crowded and inaccessible locations, to test battered piles, and to check pullout capacity as well as downward load capacity.

6. Analysis of Load Tests. Table 5-3 illustrates four methods of estimating ultimate bearing capacity of a pile from data that may be obtained from a load-displacement test such as described in ASTM D 1143. At least three of these methods, depending on local experience or preference, should be used to determine a suitable range of probable bearing capacity. The methods given in Table 5-3 give a range of ultimate pile capacities varying from 320 to 467 kips for the same pile load test data.

5-3. Capacity to Resist Uplift and Downdrag

Deep foundations may be subject to other vertical loads such as uplift and downdrag forces. Uplift forces are caused by pullout loads from structures or heave of expansive soils surrounding the shaft tending to drag the shaft up. Downdrag forces are caused by

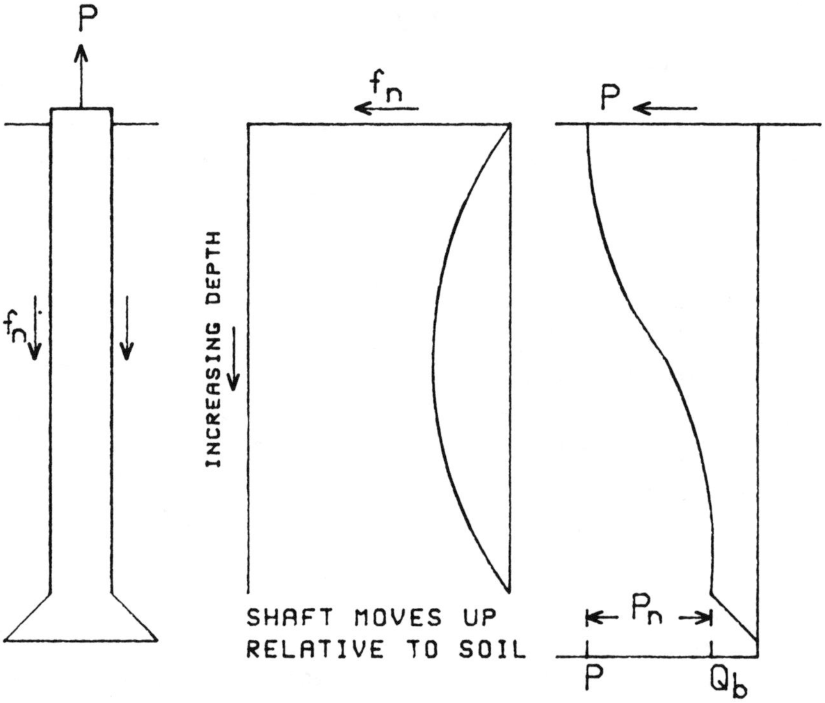

Figure 5-9. Underreamed drilled shaft resisting pullout

settlement of soil surrounding the shaft that exceeds the downward displacement of the shaft and increases the downward load on the shaft. These forces influence the skin friction that is developed between the soil and the shaft perimeter and influences bearing capacity.

A. METHOD OF ANALYSIS. Analysis of bearing capacity with respect to these vertical forces requires an estimate of the relative movement between the soil and the shaft perimeter and the location of neutral point *n*, the position along the shaft length where there is no relative movement between the soil and the shaft. In addition, tension or compression stresses in the shaft or pile caused by uplift or downdrag must be calculated to properly design the shaft. These calculations are time-dependent and complicated by soil movement. Background theory for analysis of pullout, uplift and downdrag forces of single circular drilled shafts, and a method for computer analysis of these forces is provided below. Other methods of evaluating vertical capacity for uplift and downdrag loads are given in Reese and O'Neill (1988).

B. PULLOUT. Deep foundations are frequently used as anchors to resist pullout forces. Pullout forces are caused by overturning moments such as from wind loads on tall structures, utility poles, or communication towers.

1. Force Distribution. Deep foundations may resist pullout forces by shaft skin resistance and resistance mobilized at the tip contributed by enlarged bases illustrated in Figure 5-9. The shaft resistance is defined in terms of negative skin friction f_n to indicate that the shaft is moving up relative to the soil. This is in contrast to compressive loads that are resisted by positive skin friction where the shaft moves down relative to the soil, Figure 5-4. The shaft develops a tensile stress from pullout forces. Bearing capacity resisting pullout may be estimated by

$$P_u = Q_{bu} + P_{nu} + W_p \quad (5\text{-}14a)$$

$$P_u = q_{bu}A_{bp} + \sum_{i=1}^{n} P_{nui} + W_p \quad (5\text{-}14b)$$

where

P_u = ultimate pullout resistance, kips

A_{bp} = area of base resisting pullout force, ft^2

P_{nui} = pullout skin resistance for pile element *i*, kips

2. End Bearing Resistance. Enlarged bases of drilled shafts resist pullout and uplift forces. q_{bu} may be estimated using Equation 5-2c. Base area A_b resisting pullout to be used in Equation 5-1b for

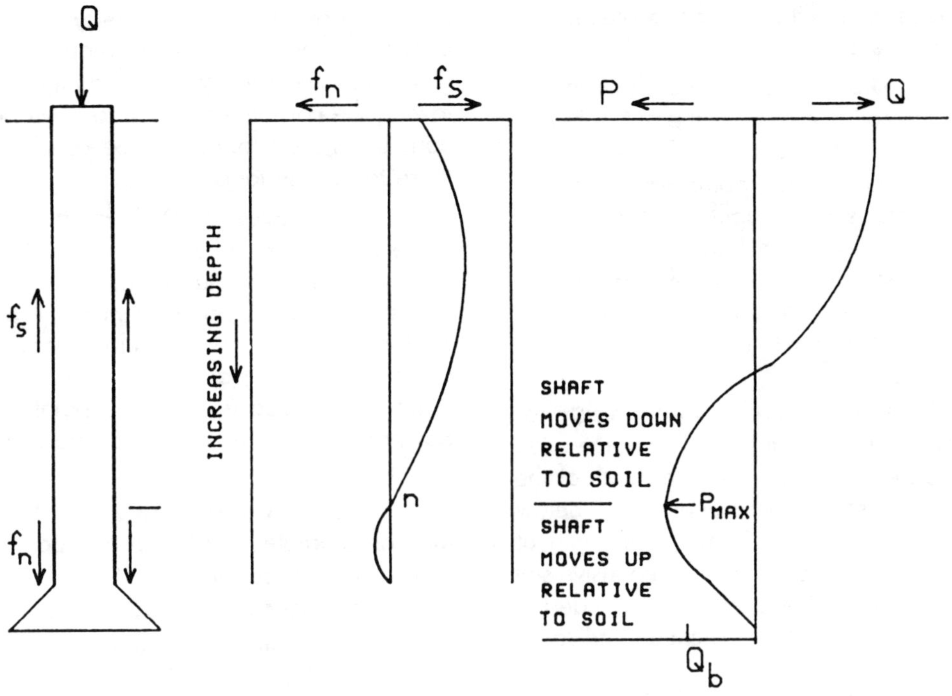

Figure 5-10. Deep foundation resisting uplift thrust

underreamed drilled shafts is

$$A_{bp} = \frac{\pi}{4} \cdot (B_b^2 - B_s^2) \qquad (5\text{-}15)$$

where

B_b = diameter of base, ft

B_s = diameter of shaft, ft

a. Cohesive Soil. The undrained shear strength C_u to be used in Equation 5-3 is the average strength above the base to a distance of $2B_b$. N_{cp} varies from 0 at the ground surface to a maximum of 9 at a depth of $2.5B_b$ below the ground (Vesic 1971).

b. Cohesionless Soil. N_{qp} of Equation 5-2 can be obtained from Equation 5-7 of the Vesic Alternate Method where ζ_{qp} is given by Equations 5-6.

3. Skin Resistance. The diameter of the shaft may be slightly reduced from pullout forces by a Poisson effect that reduces lateral earth pressure on the shaft perimeter. Skin resistance will therefore be less than that developed for shafts subject to compression loads because horizontal stress is slightly reduced. The mobilized negative skin friction f_{ni} may be estimated as 2/3 of that evaluated for compression loads f_{si}

$$\sum_{i=1}^{n} P_{nui} = C_s \cdot \Delta L \sum_{i=1}^{n} f_{ni} \qquad (5\text{-}16a)$$

$$f_{ni} = \frac{2}{3} \cdot f_{si} \qquad (5\text{-}16b)$$

where

C_s = shaft circumference, ft

ΔL = length of pile element i, ft

f_{si} = positive skin friction of element i from compressive loading using Equations 5-10 to 5-12

The sum of the elements equals the shaft length.

C. UPLIFT. Deep foundations constructed in expansive soil are subject to uplift forces caused by swelling of expansive soil adjacent to the shaft. These uplift forces cause a friction on the upper length of the shaft perimeter tending to move the shaft up. The shaft perimeter subject to uplift thrust is in the soil subject to heave. This soil is often within the top 7 to 20 ft of the soil profile referred to as the depth of the active zone for heave Z_a. The shaft located within Z_a is sometimes constructed to isolate the shaft perimeter from the expansive soil to reduce this uplift thrust. The shaft base and underream resisting uplift should be located below the depth of heaving soil.

1. Force Distribution. The shaft moves down relative to the soil above neutral point *n*, Figure 5-10, and moves up relative to the soil below point *n*. The negative skin friction f_n below point *n* and enlarged bases of drilled shafts resist the uplift thrust of expansive soil. The positive skin friction f_s above point *n* contributes to uplift thrust from heaving soil and puts the shaft in tension. End bearing and skin friction capacity resisting uplift thrust may be estimated by Equations 5-14.

2. End Bearing. End bearing resistance may be estimated similar to that for pullout forces. N_{cp} should be assumed to vary from 0 at the depth of the active zone of heaving soil to 9 at a depth $2.5B_b$ below the depth of the active zone of heave. The depth of heaving soil may be at the bottom of the expansive soil layer or it may be estimated by guidelines provided in TM 5-818-7, EM 1110-1-1904, or McKeen and Johnson (1990).

3. Skin Friction. Skin friction from the top of the shaft to the neutral point *n* contributes to uplift thrust, while skin friction from point *n* to the base contributes to skin friction that resists the uplift thrust.

a. The magnitude of skin friction f_s above point *n* that contributes to uplift thrust will be as much or greater than that estimated for compression loads. The adhesion factor α_a of Equation 5-10 can vary from 0.6 to 1.0 and can contribute to shaft heave when expansive soil is at or near the ground surface. α_a should not be underestimated when calculating the potential for uplift thrust, otherwise, tension, steel reinforcement, and shaft heave can be underestimated.

b. Skin friction resistance f_n that resists uplift thrust should be estimated similar to that for pullout loads because uplift thrust places the shaft in tension tending to pull the shaft out of the ground and may slightly reduce lateral pressure below neutral point *n*.

D. DOWNDRAG. Deep foundations constructed through compressible soils and fills can be subject to an additional downdrag force. This downdrag force is caused by the soil surrounding the drilled shaft or pile settling downward more than the deep foundation. The deep foundation is dragged downward as the soil moves down. The downward load applied to the shaft is significantly increased and can even cause a structural failure of the shaft as well as excessive settlement of the foundation. Settlement of the soil after installation of the deep foundation can be caused by the weight of overlying fill, settlement of poorly compacted fill and lowering of the groundwater level. The effects of downdrag can be reduced by isolating the shaft from the soil, use of a bituminous coating or allowing the consolidating soil to settle before construction. Downdrag loads can be considered by adding these to column loads.

1. Force Distribution. The shaft moves up relative to the soil above point *n* (Figure 5-11) and moves down relative to the soil below point *n*. The positive skin friction f_s below point *n* and end bearing capacity resists the downward loads applied to the shaft by the settling soil and the structural loads. Negative skin friction f_n above the neutral point contributes to the downdrag load and increases the compressive stress in the shaft.

2. End Bearing. End bearing capacity may be estimated similar to methods for compressive loads given by Equations 5-2.

3. Skin Friction. Skin friction may be estimated by Equation 5-9 where the positive skin friction is given by Equations 5-10 to 5-12.

E. COMPUTER ANALYSIS. Program AXILTR (AXIal Load-TRansfeR), Appendix C, is a computer program that computes the vertical shaft and soil displacements for axial down-directed structural, axial pullout, uplift, and down-drag forces using equations in Table 5-4. Load-transfer functions are used to relate base pressures and skin friction with displacements. Refer to Appendix C for example applications using AXILTR for pullout, uplift, and downdrag loads.

1. Load-Transfer Principle. Vertical loads are transferred from the top of the shaft to the supporting soil adjacent to the shaft using skin friction-load transfer functions and to soil beneath the base using consolidation theory or base load-transfer functions. The total bearing capacity of the shaft Q_u is the sum of the total skin Q_{su} and base Q_{bu} resistances given by Equations 5-1.

a. The load-displacement calculations for rapidly applied downward vertical loads have been validated by comparison with field test results (Gurtowski and Wu 1984). The strain distribution from uplift forces for drilled shafts in shrink/swell soil have been validated from results of load tests (Johnson 1984).

b. The program should be used to provide a minimum and maximum range for the load-displacement behavior of the shaft for given soil conditions. A listing of AXILTR is provided to allow users to update and calibrate this program from results of field experience.

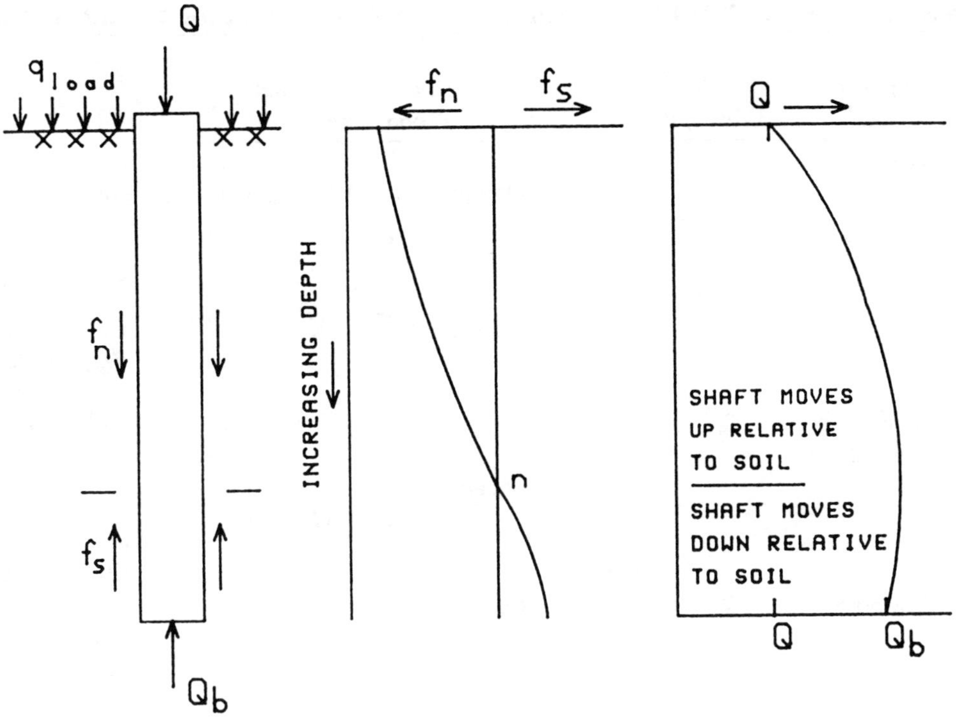

Figure 5-11. Deep foundation resisting downdrag. q_{load} is an area pressure from loads such as adjacent structures

2. Base Resistance Load Transfer. The maximum base resistance q_{bu} in Equation 5-1b is computed by AXILTR from Equation 5-2a

$$q_{bu} = cN_{cp} + \sigma_L' N_{qp} \qquad (5\text{-}17)$$

where

c = cohesion, psf

N_{cp} = cohesion bearing capacity factor, dimensionless

N_{qp} = friction bearing capacity factor, dimensionless

σ_L' = effective vertical overburden pressure at the pile base, psf

Correction factors ζ are assumed unity and the $N_{\gamma p}$ term is assumed negligible. Program AXILTR does not limit σ_L'.

a. N_{qp} for effective stress analysis is given by Equation 5-7 for local shear (Vesic Alternate method) or Equation 5-8 for general shear.

b. N_{cp} for effective stress analysis is given by

$$N_{cp} = (N_{qp} - 1)\cot\phi' \qquad (5\text{-}5a)$$

N_{cp} for total stress analysis is assumed 9 for general shear and 7 for local shear; N_{qp} and total stress friction angle ϕ are zero for total stress analysis.

c. End bearing resistance may be mobilized and base displacements computed using the Reese and Wright (1977) or Vijayvergiya (1977) base load-transfer functions (Figure 5-12a) or consolidation theory. Ultimate base displacement for the Reese and Wright model is computed by

$$\rho_{bu} = 2B_b \cdot \epsilon_{50} \cdot 12 \qquad (5\text{-}18)$$

where

ρ_{bu} = ultimate base displacement, in.

B_b = diameter of base, ft

ϵ_{50} = strain at 1/2 of maximum deviator stress of clay from undrained triaxial test, fraction

The ultimate base displacement for the Vijayvergiya model is taken as 4 percent of the base diameter.

d. Base displacement may be calculated from consolidation theory for overconsolidated soils as described in Chapter 3, Section III of EM 1110-1-1904. This calculation assumes no wetting beneath the base of the shaft from exterior water sources, except for the effect of changes in water level elevations. The calculated settlement is based on effective stresses relative to the initial effective pressure on the soil beneath the base of the shaft prior to placement of any struc-

Table 5-4. Program AXILTR Shaft Resistance To Pullout, Uplift and Downdrag Loads

Soil Volume Change	Type of Applied Load	Applied Load, Pounds	Resistance to Applied Load, Pounds	Equations
None	Pullout	$Q_{DL} - P$	Straight: $Q_{sur} + W_p$	$Q_{sur} = \pi\ B_s \int_0^L f_{\bar{s}}\ dl$
			Underream: Smaller of $Q_{sub} + W_p$ $Q_{sur} + Q_{bur} + W_p$	$Q_{sub} = \pi B_b \int_0^L \tau_s dL$ $Q_{bur\ b} = q_{bu}(B_b^2 - \cdot B_s^2)$ $W_{p\ s} = L\gamma_p B_s^2$
Swelling Soil	Uplift thrust	Q_{us}	Straight: $Q_{sur} + W_p$	$Q_{us} = \pi B_s \int_0^{L_n} f_s^{dL}\ f_{\bar{s}} dL$
			Underream: $Q_{sur} + Q_{bur} + W_p$	$Q_{sur\ s} = \pi B_s \int_{L_n}^{L} f_s dL$ $Q_{bur\ b} = q_{bu}(B_b^2 - B_s^2)$
Settling soil	Downdrag	$Q_d + Q_{sud}$	$Q_{sur} + Q_{bu}$	$Q_{sud} = \pi B_s \int_0^{L_n} f_n dL$ $q_{sur\ s} = \pi B_s \int_{L_n}^{L} f_s dL$

Nomenclature:

B_b	Base diameter, ft	Q_{DL}	Dead load of structure, pounds
B_s	Shaft diameter, ft	P	Pullout load, pounds
f_s	Maximum mobilized shear strength, psf	Q_{sub}	Ultimate soil shear resistance of cylinder diameter B_b and length equal to depth of underream, pounds
f_n	Negative skin friction, psf		
L	Shaft length, ft		
L_n	Length to neutral point n, ft	Q_{sud}	Downdrag, pounds
q_{bu}	Ultimate base resistance, psf	Q_{sur}	Ultimate skin resistance, pounds
Q_{bu}	Ultimate base capacity, pound	Q_{us}	Uplift thrust, pounds
Q_{bur}	Ultimate base resistance of upper portion of underream, pounds	Q_d	Design Load, Dead + Live loads, pounds
		W_p	Shaft weight, pounds
τ_s	Soil shear strength, psf	γ_p	Unit shaft weight, pounds/ft^3

tural loads. The effective stresses include any pressure applied to the surface of the soil adjacent to the shaft. AXILTR may calculate large settlements for small applied loads on the shaft if the maximum past pressure is less than the initial effective pressure simulating an underconsolidated soil. Effective stresses in the soil below the shaft base caused by loads in the shaft are attenuated using Boussinesq stress distribution theory (Boussinesq 1885).

3. Underream Resistance. The additional resistance provided by a bell or underream for pullout or uplift forces is $\frac{7}{9}$ of the end bearing resistance. If applied downward loads at the base of the shaft exceed the calculated end bearing capacity, AXILTR prints "THE BEARING CAPACITY IS EXCEEDED". If pullout loads exceed the pullout resistance, the program prints "SHAFT PULLS OUT". If the shaft heave exceeds the soil heave, the program prints "SHAFT UNSTABLE".

4. Skin Resistance Load Transfer. The shaft skin friction load-transfer functions applied by program AXILTR are the Seed and Reese (1957) and Kraft, Ray, and Kagawa (1981) models illustrated in Figure 5-12b. The Kraft, Ray, and Kagawa model requires an estimate of a curve fitting constant R from

$$G_s = G_i\left[1 - \frac{\tau R}{\tau_{max}}\right] \qquad (5\text{-}19a)$$

where

G_s = soil shear modulus at an applied shear stress τ, pounds/square foot (psf)

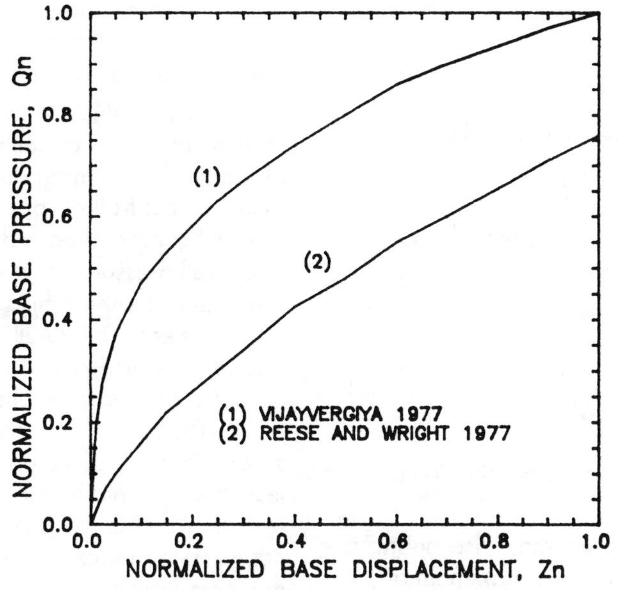

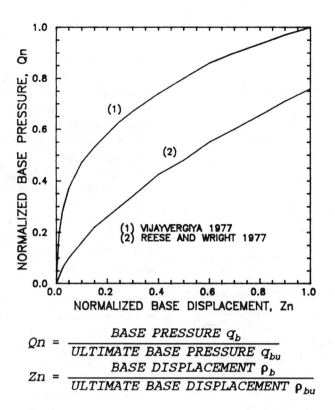

$$Qn = \frac{BASE\ PRESSURE\ q_b}{ULTIMATE\ BASE\ PRESSURE\ q_{bu}}$$

$$Zn = \frac{BASE\ DISPLACEMENT\ \rho_b}{ULTIMATE\ BASE\ DISPLACEMENT\ \rho_{bu}}$$

a. BASE TRANSFER (NORMALIZED Q–Z) FUNCTIONS

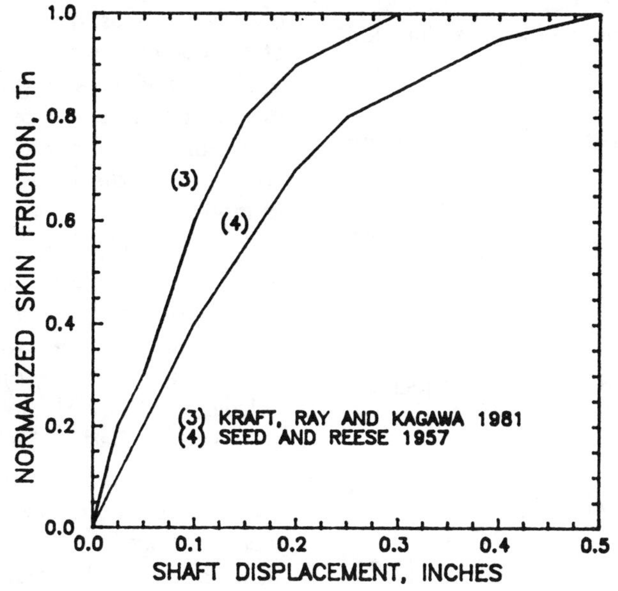

$$Tn = \frac{MOBILIZED\ SKIN\ FRICTION\ f_s}{MAXIMUM\ MOBILIZED\ SKIN\ FRICTION f_{\bar{s}}}$$

b. SHAFT TRANSFER (NORMALIZED T–Z) FUNCTIONS

Figure 5-12. Load-transfer curves applied in AXILTR

G_i = initial shear modulus, psf

τ = shear stress, psf

τ_{max} = shear stress at failure, psf

R = curve fitting constant, usually near 1.0

The curve fitting constant R is the slope of the relationship of $1 - G_s/G_i$ versus τ/τ_{max} and may be nearly 1.0. The soil shear modulus G_s is found from the elastic soil modulus E_s by

$$G_s = \frac{E_s}{2(1 + \upsilon_s)} \qquad (5\text{-}19b)$$

where υ_s is the soil Poisson's ratio. A good value for υ_s is 0.3 to 0.4.

a. Load-transfer functions may also be input into AXILTR for each soil layer up to a maximum of 11 different functions. Each load-transfer function consists of 11 data values consisting of the ratio of the mobilized skin friction/maximum mobilized skin friction $f_s/f_{\bar{s}}$ correlated with displacement as illustrated in Figure 5-12b. The maximum mobilized skin friction $f_{\bar{s}}$ is assumed the same as the maximum soil shear strength. The corresponding 11 values of the shaft displacement (or shaft movement) in inches are input only once and applicable to all of the load-transfer functions. Therefore, the values of $f_s/f_{\bar{s}}$ of each load transfer function must be correlated with the given shaft displacement data values.

b. The full mobilized skin friction $f_{\bar{s}}$ is computed for effective stresses from

$$f_{\bar{s}} = c' + \beta_f \sigma'_v \qquad (5\text{-}20)$$

where

c' = effective cohesion, psf

β_f = lateral earth pressure and friction angle factor

σ'_v = effective vertical stress, psf

The factor β_f is calculated in AXILTR by

$$\beta_f = K_o \tan\phi' \qquad (5\text{-}21)$$

where

K_o = lateral coefficient of earth pressure at rest

ϕ' = effective peak friction angle from triaxial tests, deg

The effective cohesion is usually ignored.

c. The maximum mobilized skin friction $f_{\bar{s}}$ for each element is computed for total stresses from Equation 5-10 using α_a from Table 5-1 or Equations 5-11.

5. Influence of Soil Movement. Soil movement, heave or settlement, alters the performance of the shaft. The magnitude of the soil heave or settlement is controlled by the swell or recompression indices, compression indices, maximum past pressure and swell pressure of each soil layer, depth to the water table, and depth of the soil considered active for swell or settlement. The swell index is the slope of the rebound pressure-void ratio curve on a semi-log plot of consolidation test results as described in ASTM D 4546. The recompression index is the slope of the pressure-void ratio curve on a semi-log plot for pressures less than the maximum past pressure. The swell index is assumed identical with the recompression index. The compression index is the slope of the linear portion of the pressure-void ratio curve on a semi-log plot for pressures exceeding the maximum past pressure. The maximum past pressure (preconsolidation stress) is the greatest effective pressure to which a soil has been subjected. Swell pressure is defined as a pressure which prevents a soil from swelling at a given initial void ratio as described by method C in ASTM D 4546.

a. The magnitude of soil movement is determined by the difference between the initial and final effective stresses in the soil layers and the soil parameters. The final effective stress in the soil is assumed equivalent with the magnitude of the total vertical overburden pressure, an assumption consistent with zero final pore water pressure. Program AXILTR does not calculate soil displacements for shaft load transferred to the soil.

b. Swell or settlement occurs depending on the difference between the input initial void ratio and the final void ratio determined from the swell and compression indices, the swell pressure, and the final effective stress for each soil element. The method used to calculate soil swell or settlement of soil adjacent to the shaft is described as Method C of ASTM D 4546.

c. The depth of the active zone Z_a is required and it is defined as the soil depth above which significant changes in water content and soil movement can occur because of climate and environmental changes after construction of the foundation. Refer to EM 1110-1-1904 for further information.

5-4. Lateral Load Capacity of Single Shafts

Deep foundations may be subject to lateral loads as well as axial loads. Lateral loads often come

Table 5-5. Brom's Equations for Ultimate Lateral Load (Broms 1964a, Broms 1964b, Broms 1965)

a. Free Head Pile in Cohesive Soil

Pile	Equations		Diagram
Short $L \le L_c$	$T_u = 18C_uB_s\left[(e^2+1.5B_se+eL+0.5L^2+1.125B_s^2)^{1/2}-(e+0.75B_s+0.5L)\right]$	(5-22a)	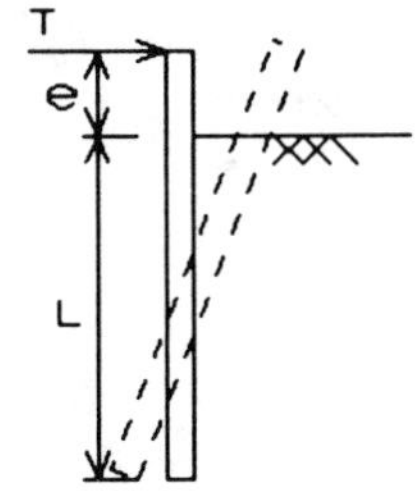
	$L_c = 1.5B_s + \frac{9}{C_uB_s} + \left[\frac{M_y}{2.25C_uB_s}\right]^{1/2}$	(5-22b)	
Long $L \ge L_c$	$T_u = 9C_uB_s\left(\left[(e+1.5B_s)^2 + \frac{2M_y}{9C_uB_s}\right]^{1/2} - e - 1.5B_s\right)$	(5-22c)	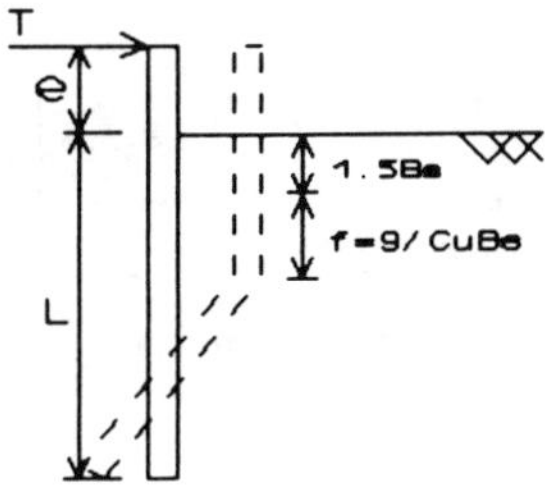
	Circular: $M_y = 1.3f_yZ$ *H-section*: $= 1.1f_yZ_{max}$ *H-section*: $= 1.5f_yZ_{min}$		

b. Free Head Pile in Cohesionless Soil

Pile	Equations		Diagram
Short $L \le L_c$	$T_{us} = \frac{\gamma B_sK_pL^3 - 2M_a}{2(e+L)}$	(5-24a)	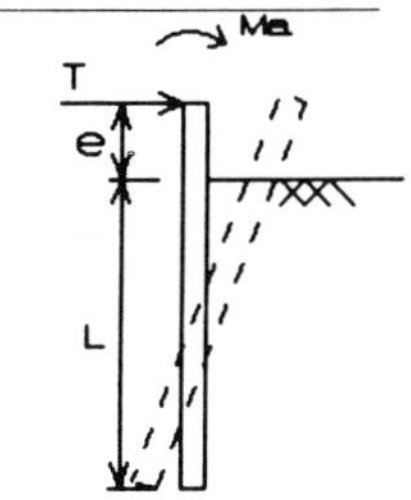
	$L_c^3 - \frac{2T_{ul}}{\gamma B_sK_p}\cdot L_c - \frac{2(M_a + T_{ul}e)}{\gamma B_sK_p} = 0$	(5-24b)	
Long $L \ge L_c$	$T_{ul} = \frac{M_y - M_a}{e + \frac{2}{3}\cdot f}$	(5-24c)	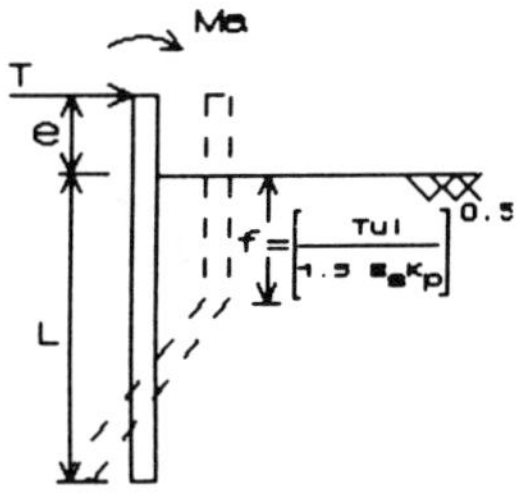

from wind forces on the structure or inertia forces from traffic. Lateral load resistance of deep foundations is determined by the lateral resistance of adjacent soil and bending moment resistance of the foundation shaft. The ultimate lateral resistance T_u often develops at lateral displacements much greater than can be allowed by the structure. An allowable lateral load T_a should be determined to be sure that the foundation will be safe with respect to failure.

A. ULTIMATE LATERAL LOAD. Brom's equations given in Table 5-5 can give good results for

Table 5-5. (Continued)

c. Fixed Head Pile in Cohesive Soil

Pile	Equations		Diagram
Short $L \le L_{cs}$	$T_u = 9C_uB_s(L - 1.5B_s)$	(5-23a)	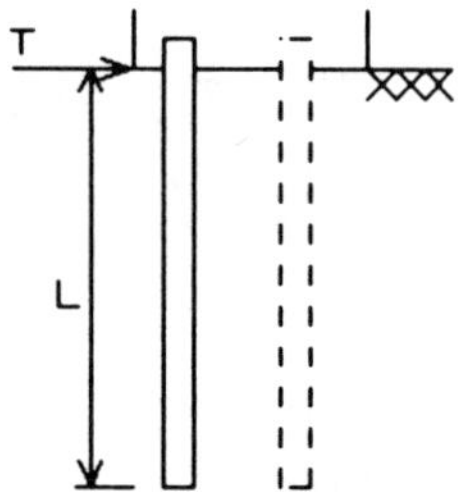
	$L_{cs} = 2\left[\frac{M_y}{18C_uB_s} + \frac{9}{16}\cdot B_s^2\right]$	(5-23b)	
Inter-mediate $L_{cs} \le L$ $L \ge L_{cl}$	$T_u = 18C_uB_s\left[\frac{M_y}{9C_uB_s} + \frac{L^2}{2} + \frac{9}{8}\cdot B_s^2\right]^{1/2} - (0.75B_s + 0.5L)$	(5-23c)	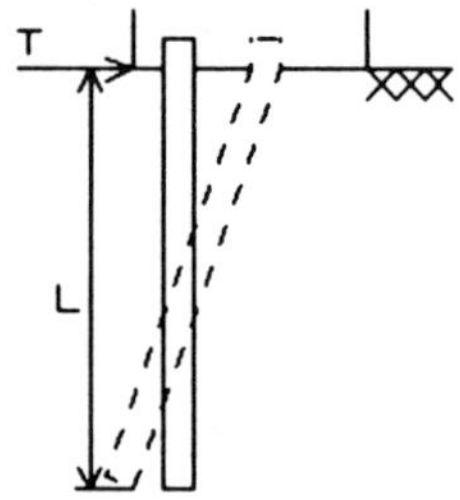
	$L_{cl} = \left[2.25B_s^2 + \frac{4}{9}\cdot\frac{M_y}{9C_uB_s}\right]^{1/2} + \left[\frac{M_y}{2.25C_uB_s}\right]^{1/2}$	(5-23d)	
Long $L \ge L_{cl}$	$T_u = 9C_uB_s\left[(2.25B_s^2 + \frac{4}{9}M_y)^{1/2} - 1.5B_s\right]$	(5-23e)	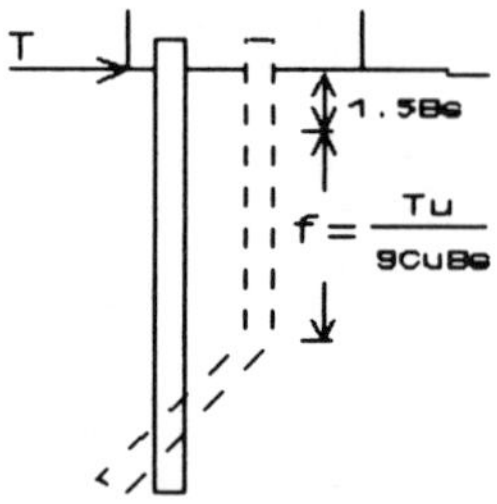

d. Fixed Head Pile in Cohesionless Soil

Pile	Equations		Diagram
Short $L \le L_{cs}$	$T_s = 1.5\gamma B_sK_pL^2$	(5-25a)	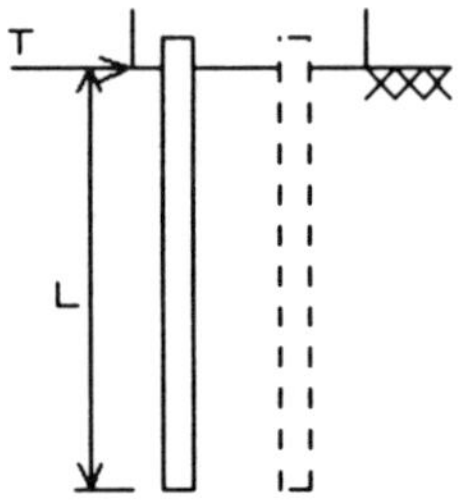
	$L_{cs} = \left[\frac{M_y}{\gamma B_sK_p}\right]^{1/3}$	(5-25b)	

many situations and these are recommended for an initial estimate of ultimate lateral load T_u. Ultimate lateral loads can be readily determined for complicated soil conditions using a computer program such as COM624G based on beam-column theory and given *p-y* curves (Reese, Cooley, and Radhakkrishnan 1984). A *p-y* curve is the relationship between the soil resistance per shaft length (kips/inch) and the deflection (inches) for a given lateral load *T*.

1. Considerations.

a. Lateral load failure may occur in short drilled shafts and piles, which behave as rigid members, by soil failure and excessive pile deflection and in long piles by excessive bending moment.

b. Computation of lateral deflection for different shaft penetrations may be made to determine the depth of penetration at which additional penetration

Table 5-5. (Concluded)

Pile	Equations		Diagram
Inter-mediate $L_{cs} \le L_{cl}$ $L \ge L_{cl}$	$T_u = \frac{M_y}{L} + 0.5\gamma B_s K_p$	(5-25c)	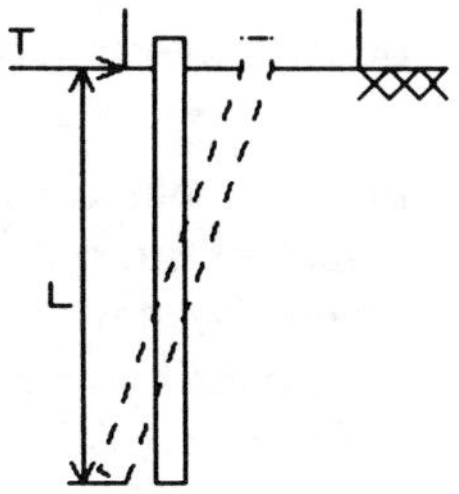
	$L_{cl}^3 - \frac{T_{ul}}{0.5\gamma B_s K_p} \cdot L_{cl} + \frac{M_y}{0.5\gamma B_s K_p}$	(5-25d)	
Long $L \ge L_{cl}$	$T_u = \frac{2M_y}{e + \frac{2}{3} \cdot f}$	(5-25e)	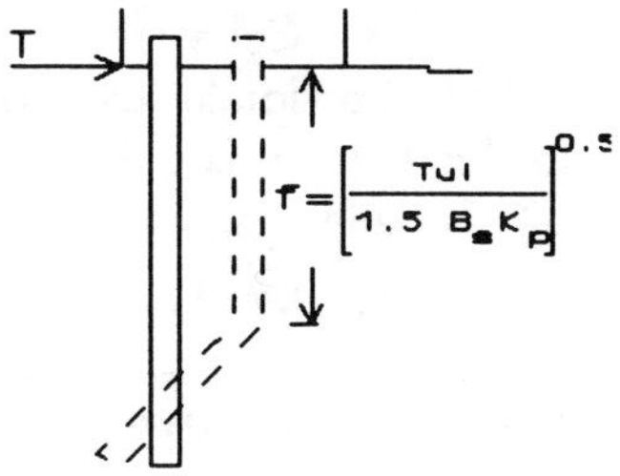

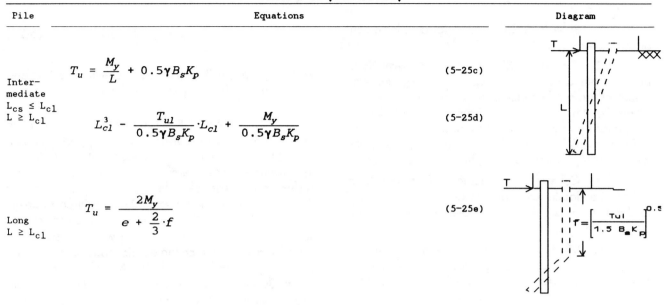

e. Nomenclature

B_s	= diameter of pile shaft, ft
C_u	= undrained shear strength, kips/ft^2
c	= distance from centroid to outer fiber, ft
e	= length of pile above ground surface, ft
$1.5B_s + f$	= distance below ground surface to point of maximum bending moment in cohesive soil, ft
f	= distance below ground surface at point of maximum bending moment in cohesionless soil, ft
f_y	= pile yield strength, ksf
I_p	= pile moment of inertia, ft^4
K_p	= Rankine coefficient of passive pressure, $\tan^2(45 + \phi'/2)$
L	= embeded pile length, ft
L_c	= critical length between long and short pile, ft
L_{cs}	= critical length between short and intermediate pile, ft
L_{cl}	= critical length between intermediate and long pile, ft
M_a	= applied bending moment, positive in clockwise direction, kips-ft
M_y	= ultimate resisting bending moment of entire cross-section, kips-ft
T	= lateral load, kips
T_u	= ultimate lateral load, kips
T_{ul}	= ultimate lateral load of long pile in cohesionless soil, kips
Z	= section modulus I_p/c, ft^3
Z_{max}	= maximum section modulus, ft^3
Z_{min}	= minimum section modulus, ft^3
γ	= unit wet weight of soil, kips/ft^3
ϕ'	= effective angle of internal friction of soil, degrees

will not significantly decrease the groundline deflection. This depth will be approximately $4\beta_\ell$ for a soil in which the soil stiffness increases linearly with depth

$$\beta_\ell = \left[\frac{E_p I_p}{k}\right]^{1/5} \quad (5\text{-}26)$$

where

E_p = elastic modulus of shaft or pile, ksf

I_p = moment of inertia of shaft, ft^4

k = constant relating elastic soil modulus with depth, $E_s = kz$ kips/ft^3

Shafts which carry insignificant axial loads such as those supporting overhead signs can be placed at this minimum depth if their lateral load capacity is acceptable.

c. Cyclic loads reduce the support provided by the soil, cause gaps to appear near the ground surface adjacent to the shaft, increase the lateral deflection for a given applied lateral load and can reduce the ultimate lateral load capacity because of the loss of soil support.

d. Refer to ASTM D 3966 for details on conducting lateral load tests.

2. Broms' Closed Form Solution. Broms' method uses the concept of a horizontal coefficient of subgrade reaction and considers short pile failure by flow of soil around the pile and failure of long piles by forming a plastic hinge in the pile. Refer to Broms (1964a), Broms (1964b), Broms (1965), and Reese (1986) for estimating T_u from charts.

a. Cohesive soil to depth $1.5B_s$ is considered to have negligible resistance because a wedge of soil to depth $1.5B_s$ is assumed to move up and when the pile is deflected.

b. Iteration is required to determine the ultimate lateral capacity of long piles T_{ul} in cohesionless soil, Table 5-5. Distance *f*, Table 5-5b and 5-5d, may first be estimated and T_{ul} calculated; then, *f* is calculated and T_{ul} recalculated as necessary. T_{ul} is independent of length *L* in long piles.

3. Load Transfer Analysis. The method of solution using load transfer *p-y* curves is also based on the concept of a coefficient of horizontal subgrade reaction. A fourth-order differential equation is solved using finite differences and load transfer *p-y* curves.

a. Numerous *p-y* relationships are available to estimate appropriate values of soil stiffness for particular soil conditions (Reese 1986). *p-y* curves developed from local experience and back-calculated from lateral load tests may also be used in program COM624G.

b. Program COM624G has provided excellent agreement with experimental data for many load test results.

B. ALLOWABLE LATERAL LOADS. Estimates of allowable lateral load T_a is best accomplished from results of lateral load-deflection (*p-y*) analysis using given *p-y* cuves and a computer program such as COM624G. The specified maximum allowable lateral deflection should be used to estimate T_a.

1. Minimum and maximum values of the expected soil modulus of subgrade reaction should be used to determine a probable range of lateral load capacity. This modulus may be estimated from results of pressuremeter tests using the Menard deformation modulus (Reese 1986), estimates of the elastic soil modulus with depth, or values given in Table 5-6b.

2. A rough estimate of allowable lateral load T_a may be made by calculating lateral groundline deflection y_o using Equations in Table 5-6,

$$T_a = \frac{y_a}{y_o} \cdot T_u \qquad (5\text{-}27)$$

where y_a is a specified allowable lateral deflection and T_u is estimated from equations in Table 5-5.

C. EXAMPLE APPLICATION. A concrete drilled shaft is to be constructed to support a design lateral load T_d = 10 kips. This load will be applied at the ground surface, therefore length above the ground surface $e = 0$. Lateral deflection should be no greater than y_a = 0.25 inch. An estimate is required to determine a suitable depth of penetration and diameter to support this lateral load in a clay with cohesion C_u = 1 ksf for a soil in which the elastic modulus is assumed to increase linearly with depth. A trial diameter B_s = 2.5 ft (30 inches) is selected with 1 percent steel. Yield strength of the steel f'_{ys} = 60 ksi and concrete strength f'_c = 3 ksi.

1. Minimum Penetration Depth. The minimum penetration depth may be estimated from Equation 5-26 using E_pI_p and *k*. Table 5-7 illustrates calculation of E_pI_p for a reinforced concrete shaft which is $2.7 \cdot 10^5$ kips-ft^2. k = 170 kips/ft^3 from Table 5-6b when the elastic modulus increases linearly with depth. Therefore,

$$\beta_\ell = \left[\frac{E_pI_p}{k}\right]^{1/5} = \left[\frac{2.7 \cdot 10^5}{170}\right]^{1/5} = 4.37 \text{ ft}$$

The minimum depth of penetration $L = 4\beta_\ell = 4 \cdot 4.37 =$ 17.5 ft. Select L = 20 ft.

2. Ultimate Lateral Load. Broms equations in Table 5-5a for a free head pile in cohesive soil may be used to roughly estimate T_u. The ultimate bending moment resistance M_y using data in Table 5-7 is

$$M_y = 1.3f_yZ = 1.3[f'_{ys}Z_{st} + 0.03f'_cZ_c]$$

$$= 1.3\left[f'_{ys}\sum\frac{I_{st}}{c_{st}} + 0.03f'_c\frac{2I_g}{B_s}\right]$$

$$= 1.3\left[60 \cdot 2 \cdot 0.79(11.33 + 9.96 + 7.39 + 3.93)\right.$$

$$\left. + 0.03 \cdot 3\,\frac{2 \cdot 39760}{30}\right]$$

$$= 1.3\,[94.8 \cdot 32.61 + 238.56] = 4329 \text{ kip-in}$$

Table 5-6. Estimation of Ultimate Lateral Deflection y_o at the Groundline (Broms 1964a, Reese 1986)

a. Soil With Modulus of Subgrade Reaction Constant With Depth

Pile	Equation	Remarks
Short Free Head $\beta_c L < 1.5$	$y_o = \dfrac{4T_u\left(1 + 1.5\dfrac{e}{L}\right)}{E_{si}\,L}$	$\beta_c = \left[\dfrac{E_{sl}}{4E_p I_p}\right]^{1/4}$
Short Fixed $\beta_c L < 0.5$	$y_o = \dfrac{T_u}{E_{sl}L}$	E_p = pile lateral elastic modulus, ksf I_p = pile moment of inertia, ft^4 $E;_{sl}$ = modulus of subgrade reaction, *ksf*
Long Free Head $\beta_c L > 1.5$	$y_o = \dfrac{2T_u\beta_c}{E_{sl}}$	Terzaghi Recommendations for E_{sl}
Long Fixed Head $\beta_c L > 1.5$	$y_o = \dfrac{T_u\beta_c}{E_{sl}}$	

Terzaghi Recommendations for E_{sl}

Clay	C_u, ksf	E_{sl}, ksf
Stiff	1–2	3–6
Very Stiff	2–4	6–13
Hard	> 4	> 13

b. Soil With Modulus of Subgrade Reaction Increasing Linearly With Depth

Equation	Definitions
$y_o = F_y \dfrac{T_u\,\beta_\ell^3}{E_p I_p}$	$\beta_\ell = \left[\dfrac{E_p I_p}{k}\right]^{1/5}$

k = constant relating elastic soil modulus with depth, $E_s = kz$, kips/ft^3

Representative Values for k

	k, kips/ft^3	
C_u, kips/ft^2	Static	Cyclic
0.25–0.5	50	20
0.50–1.0	170	70
1.0 –2.0	500	200
2.0 –4.0	1700	700
4.0 –8.0	5000	2000

Values for F_y

$\dfrac{L}{\beta_\ell}$	F_y
2	1.13
3	1.03
4	0.96
5	0.93

Table 5-7. Example E_pI_p Computation of Drilled Shafts

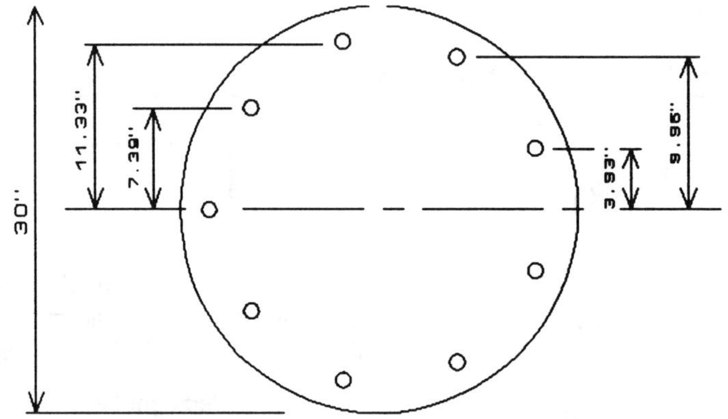

Cross-section area:	707 in^2
Steel area (1 %):	7.07 in^2 < 7.11 in^2 for 9 #8 bars, ASTM 60 grade steel
ASTM 60 grade steel	f'_{ys} = 60,000 psi
Concrete strength	f'_c = 3,000 psi
$E_c = 57.5\cdot(f'_c)^{1/2}$:	3149 kips/in^2

Gross moment of inertia:

$I_g = \pi B_s^4/64 = \pi\cdot 30^4/64 = 39{,}760 \text{ in}^4$

$E_{st} = 29{,}000 \text{ kips/in}^2$

Area of #8 bar, $A_{st} = 0.79 \text{ in}^2$

Steel moment of inertia about centroid axis, I_{st}:

$$I_{st} = 2\cdot A_{st}\ \Sigma\ (\text{distance from central axis})^2$$
$$= 2\cdot 0.79\cdot(11.33^2 + 9.96^2 + 7.39^2 + 3.93^2)$$
$$= 470.25 \text{ in}^4$$

Calculation of E_pI_p:

Using ACI Code Equation 10.8 (approximate)

$$E_pI_p = \frac{E_cI_g}{2.5} = \frac{3149\cdot 39760}{2.5} = 5\cdot 10^7 \ kips\text{-}in^2$$

Using ACI Code Equation 10.7 (more accurate)

$$E_pI_p = \frac{E_cI_g}{5} + E_{st}I_{st}$$
$$= 2.5\cdot 10^7 + 29000\cdot 470.25$$
$$= 3.86\cdot 10^7 \ kips\text{-}in^2$$
$$= 2.68\cdot 10^5 \ kips\text{-}ft^2$$

or 360.7 kip-ft. From Table 5-6a

$$L_c = 1.5B_s + \frac{9}{C_u B_s} + \left[\frac{M_y}{2.25 C_u B_s}\right]^{1/2}$$

$$= 1.5 \cdot 2.5 + \frac{9}{1 \cdot 2.5} + \left[\frac{360.7}{2.25 \cdot 1 \cdot 2.5}\right]^{1/2}$$

$$= 3.75 + 3.6 + 8.0 = 15.4 \text{ ft}$$

This shaft with $L = 20$ ft is considered long. From Equation 5-27c, the ultimate lateral load T_u is

$$T_u = 9C_{uB\ell}\left(\left[(1.5B_s)^2 + \frac{2M_y}{9C_u B_s}\right]^{1/2} - 1.5B_s\right)$$

$$= 9 \cdot 1 \cdot 2.5\left(\left[(1.5 \cdot 2.5)^{1/2} + \frac{2 \cdot 360.7}{9 \cdot 1 \cdot 2.5}\right)^{1/2} - 1.5 \cdot 2.5\right)$$

$$= 22.5([14.06 + 32.06]^{1/2} - 3.75) = 68.4 \text{ kips}$$

3. Allowable Lateral Load. From Table 5-6b, the ultimate lateral deflection y_o is

$$y_o = F_y \frac{T_u B_\ell^3}{E_p I_p}$$

$$= 0.95 \frac{68.4 \cdot 4.4^3}{2.7 \cdot 10^5}$$

$$= 0.022 \text{ ft}$$

or 0.26 inch. On the basis of Equation 5-27, the design displacement will be (10/68.4) · 0.26 or 0.04 inch, which is less than the specified allowable deflection $y_a = 0.25$ inch. The trial dimensions are expected to be fully adequate to support the design lateral load of 10 kips. Additional analysis using COM624G should be performed to complete a more economical and reliable design.

5-5. Capacity of Shaft Groups

Drilled shafts are often not placed in closely spaced groups because these foundations can be constructed with large diameters and can extend to deep depths. The vertical and lateral load capacities of shaft foundations are often the sum of the individual drilled shafts. The FS for groups should be 3.

A. AXIAL CAPACITY. The axial capacity of drilled shafts spaced $\geq 8B_s$ will be the sum of the capacities of individual shafts. If drilled shafts are constructed in closely spaced groups where spacing between shafts is $< 8B_s$, then the capacity of the group may be less than the sum of the capacities of the individual shafts. This is because excavation of a hole for a shaft reduces effective stresses against both the sides and bases of shafts already in place. Deep foundations where spacings between individual piles are less than 8 times the shaft width B also cause interaction effects between adjacent shafts from overlapping of stress zones in the soil, Figure 5-13. In situ soil stresses from shaft loads are applied over a much larger area leading to greater settlement and bearing failure at lower total loads.

1. Cohesive Soil. Group capacity may be estimated by efficiency and equivalent methods. The efficiency method is recommended when the group cap is isolated from the soil surface, while the equivalent method is recommended when the cap is resting on the soil surface. The equivalent method is useful for spacings $\leq 3B_s$ where B_s is the shaft or pile diameter, Figure 5-14.

a. Group ultimate capacity by the efficiency method is

$$Q_{ug} = n \cdot E_g \cdot Q_u \qquad (5\text{-}28a)$$

where

Q_{ug} = group capacity, kips
n = number of shafts in the group
E_g = efficiency
Q_u = ultimate capacity of the single shaft, kips

E_g should be > 0.7 for spacings > $3B_s$ and 1.0 for spacings > $8B_s$. E_g should vary linearly for spacings between $3B_s$ and $8B_s$. $E_g = 0.7$ for spacings $\leq 2.5B_s$.

b. Group capacity by the equivalent method is

$$Q_{ug} = 2L(B + W)C_{ua} + 9 \cdot C_{ub} \cdot B \cdot W \qquad (5\text{-}28b)$$

where

L = depth of penetration, ft
W = horizontal length of group, ft
B = horizontal width of group, ft
C_{ua} = average undrained shear strength of the cohesive soil in which the group is placed, ksf

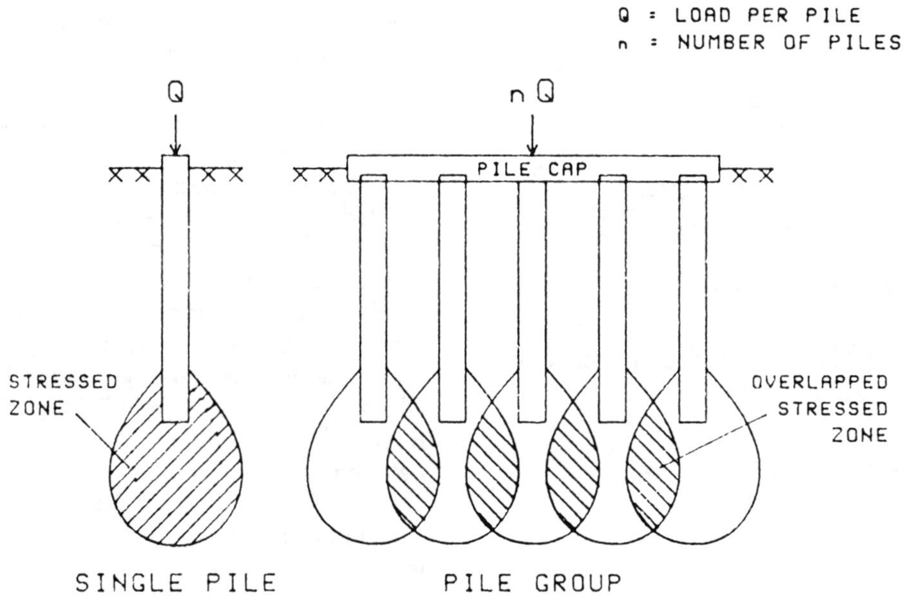

Figure 5-13. Stress zones in soil supporting group

C_{ub} = average undrained shear strength of the cohesive soil below the tip to a depth of $2B$ below the tip, ksf

The presence of locally soft soil should be checked because this soil may cause some shafts to fail.

c. The ultimate bearing capacity of a group in a strong clay soil overlying weak clay may be estimated by assuming block punching through the weak underlying soil layer. Group capacity may be calculated by Equation 5-28b using the undrained strength C_{ub} of the underlying weak clay. A less conservative solution is provided by (Reese and O'Neill 1988)

$$Q_{ug} = Q_{ug,lower} + \frac{H_b}{10B}[Q_{ug,upper} - Q_{ug,lower}] \leq Q_{ug,upper} \quad (5\text{-}29)$$

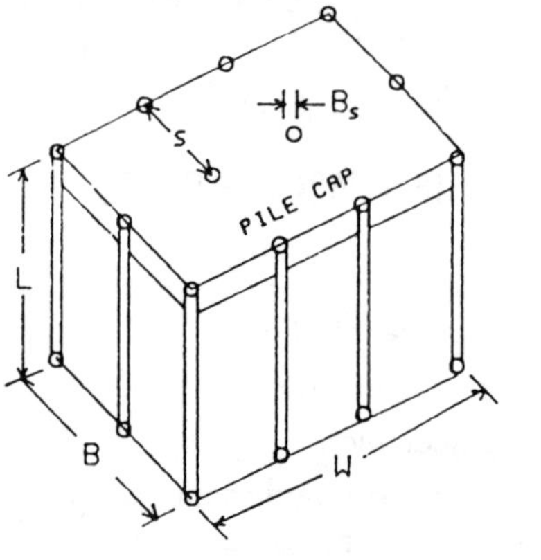

Figure 5-14. Schematic of group

where

$Q_{ug,lower}$ = bearing capacity if base at top of lower (weak) soil, kips

$Q_{ug,upper}$ = bearing capacity in the upper soil if the softer lower soil were not present, kips

H_b = vertical distance from the base of the shafts in the group to the top of the weak layer, ft

B = least width of group, ft

2. Cohesionless Soil. During construction of drilled shafts in cohesionless soil, stress relief may be more severe than in cohesive soils because cohesionless soils do not support negative pore pressures as well as cohesive soils. Negative pore pressures generated during excavation in cohesive soils tend to keep effective stresses higher than in cohesionless soil.

a. The efficiency Equation 5-28a is usually recommended.

b. Equation 5-29 can be used to estimate ultimate bearing capacity of a group in a strong cohesionless soil overlying a weak cohesive layer.

B. LATERAL LOAD CAPACITY. Response of groups to lateral load requires lateral and axial load soil-structure interaction analysis with assistance of a finite element computer program.

1. Widely-Spaced Drilled Shafts. Shafts spaced > $7B_s$ or far enough apart that stress transfer is minimal and loading is by shear, the ultimate lateral load of the group T_{ug} is the sum of individual shafts. The capacity of each shaft may be estimated by methodology in 5-4.

2. Closely-Spaced Drilled Shafts. The solution of ultimate lateral load capacity of closely spaced shafts in a group requires analysis of a nonlinear soil-shaft system

$$T_{ug} = \sum_{j=1}^{n} T_{uj} \quad (5\text{-}30)$$

where

T_{uj} = ultimate lateral load capacity of shaft j, kips

n = number of shafts in the group

Refer to Poulos (1971a, Poulos (1971b), and Reese (1986) for detailed solution of the lateral load capacity of each shaft by the Poulos-Focht-Koch method.

3. Group Behavior as a Single Drilled Shaft. A pile group may be simulated as a single shaft with diameter C_g/π where C_g is the circumference given as the minimum length of a line that can enclose the group. The moment of inertia of the group is $n \cdot I_p$ where I_p is the moment of inertia of a single shaft. Program COM624G may be used to evaluate lateral load-deflection behavior of the simulated single shaft for given soil conditions. Comparison of results between the Poulos-Focht-Koch and simulated single pile methods was found to be good (Reese 1986).

SECTION II. DRIVEN PILES

5-6. Effects of Pile Driving

Driving piles disturbs and substantially remolds soil. Driving radially compresses cohesive soils and increases the relative density of cohesionless soils near the pile.

A. COHESIVE SOIL. Soil disturbance around piles driven into soft or normally consolidated clays is within one pile diameter. Driving into saturated stiff clays closes fissures and causes complete loss of stress history near the pile (Vesic 1969).

1. Driving in Saturated Clay. Soil disturbance and radial compression increase pore water pressures that temporarily reduce the soil shear strength and pile load capacity. Pore pressures decrease with time after driving and lead to an increase in shear strength and pile load capacity. This effect is soil freeze.

2. Driving in Unsaturated Clay. Driving in unsaturated clay does not generate high pore pressures and probably will not lead to soil freeze.

B. COHESIONLESS SOIL. The load capacity of cohesionless soil depends strongly on relative density. Driving increases relative density and lateral displacement within a zone around the pile of one to two pile diameters. Large displacement piles such as closed-end pipe piles cause larger increases in relative density than small displacement piles such as H-piles. The increase in bearing capacity can therefore be greater with large displacement piles.

1. Driving in Dense Sand and Gravel. Driving in dense sand and gravel can decrease pore pressures from soil dilation and temporarily increase soil shear strength and pile load capacity. Shear strength can increase substantially and may exceed the capacity of pile driving equipment to further drive the piles into the soil. Pore pressures increase after driving and cause the shear strength to decrease and reduce the pile load capacity. This effect is soil relaxation.

2. Driving in Saturated Cohesionless Silts. Driving in saturated, cohesionless silts increases pore pressures and can temporarily reduce the soil shear strength and pile load capacity. Pore pressures dissipate after driving and lead to an increase in shear strength and pile load capacity. This effect is soil freeze as described in cohesive soil, but can occur more quickly than in cohesive soil because permeability is greater in silts.

5-7. Vertical Capacity of Single Driven Piles

The vertical capacity of driven piles may be estimated using Equations 5-1 similar to drilled shafts

$$Q_u \approx Q_{bu} + Q_{su} - W_p \quad (5\text{-}1a)$$

$$Q_u \approx q_{bu}A_b + \sum_{i=1}^{n} Q_{sui} - W_p \quad (5\text{-}1b)$$

where

Q_u = ultimate bearing capacity, kips

Q_{bu} = ultimate end bearing resistance, kips

Q_{su} = ultimate skin friction, kips

q_{bu} = unit ultimate end bearing resistance, ksf

A_b = area of tip or base, ft^2

n = number of pile elements

Q_{sui} = ultimate skin friction of pile element i, ksf

W_p = pile weight, $\approx A_b \cdot L \cdot \gamma_p$ without enlarged base, kips

L = pile length, ft

γ_p = pile density, kips/ft^3

In addition, a wave equation analysis should be performed to estimate the driving forces to prevent pile damage during driving, to estimate the total driving resistance that will be encountered by the pile to assist in determining the required capability of the driving equipment and to establish pile driving criteria. Refer to program GRLWEAP (Goble Rausche Likins and Associates, Inc. 1988) for details of wave equation analysis. Pile driving formulas are also recommended to quickly estimate the ultimate bearing capacity.

A. END BEARING CAPACITY. End bearing capacity may be estimated by

$$q_{bu} = c \cdot N_{cp} \cdot \zeta_{cp} + \sigma'_L \cdot N_{qp}\zeta_{qp} + \frac{B_b}{2} \cdot \gamma'_b \cdot N_{\gamma p} \cdot \zeta_{\gamma p} \quad (5\text{-}2a)$$

where

c = cohesion of soil beneath the tip, ksf

σ'_L = effective soil vertical overburden pressure at pile base $\approx \gamma' \cdot L$, ksf

γ'_L = effective wet unit weight of soil along shaft length L, kips/ft^3

B_b = base diameter, ft

γ'_b = effective wet unit weight of soil beneath base, kips/ft^3

$N_{cp}, N_{qp}, N_{\gamma p}$ = pile bearing capacity factors of cohesion, surcharge, and wedge components

$\zeta_{cp}, \zeta_{qp}, \zeta_{\gamma p}$ = pile soil and geometry correction factors of cohesion, surcharge, and wedge components

Equation 5-2a may be simplified for driven piles by eliminating the $N_{\gamma p}$ term

$$q_{bu} = c \cdot N_{cp} \cdot \zeta_{cp} + \sigma'_L \cdot (N_{qp} - 1) \cdot \zeta_{qp} \quad (5\text{-}2b)$$

or

$$q_{bu} = c \cdot N_{cp} \cdot \zeta_{cp} + \sigma'_L \cdot N_{qp} \cdot \zeta_{qp} \quad (5\text{-}2c)$$

Equations 5-2b and 5-2c also adjust for pile weight W_p assuming $\gamma_p \approx \gamma'_L$. Equation 5-2c is usually used because omitting the "1" is usually negligible. Bearing capacity does not increase without limit with increasing depth. Refer to Figure 5-3 to determine the critical depth Lc below which effective stress remains constant using the Meyerhof and Nordlund methods.

1. Cohesive Soil. The shear strength of cohesive soil is $c = C_u$, the undrained strength, and the effective friction angle $\phi' = 0$. Equation 5-2a leads to

$$q_{bu} = N_{cp} \cdot C_u = 9 \cdot C_u \quad (5\text{-}2d)$$

where shape factor $\zeta_{cp} = 1$ and $N_{cp} = 9$. Undrained shear strength C_u is estimated by methods in Chapter 3 and may be taken as the average shear strength within $2B_b$ beneath the tip of the pile.

2. Cohesionless Soil. Meyerhof, Nordlund, and in situ methods described below and Hanson, Vesic, and general shear methods described in Section I are recommended for solution of ultimate end bearing capacity using Equations 5-2. Several of these methods should be used for each design problem to provide a reasonable range of probable bearing capacity if calculations indicate a significant difference between methods.

a. Meyerhof Method. Figure 5-15 illustrates the bearing capacity factors to be used with Equation 5-2b (Meyerhof 1976). The range between "low" and "high" factors in Figure 5-15 should account for soil conditions such as loose or dense sands, overconsolidation ratio of clays, and soils with different degrees of compressibility. The correction factors ζ_{cp} and ζ_{qp} in Equation 5-2b are unity. N_{cp} and N_{qp} are estimated as follows:

Evaluate the critical depth ratio $Rc = Lc/B$ from the given friction angle ϕ' using Figure 5-3. Then calculate the critical depth $Lc = Rc \cdot B$ where B = pile diameter or width.

1. If $\phi' < 30°$ and $L > Lc/2$, then use $N_{cp,high}$ and $N_{qp,high}$ directly from curves of Figure 5-15

2. If $\phi' < 30°$ and $L < Lc/2$, then from Figure 5-15

$$N_{cp} = N_{cp,low} + (N_{cp,high} - N_{cp,low}) \frac{2L}{Lc} \quad (5\text{-}31a)$$

$$N_{qp} = N_{qp,low} + (N_{qp,high} - N_{qp,low}) \frac{2L}{Lc} \quad (5\text{-}31b)$$

If $\phi' \geq 30°$, evaluate the bearing depth ratio $Rb = L/B$, locate the intersection of Rb and ϕ' in Figure 5-15, then estimate by interpolation N_{cb} and N_{qp}, respectively.

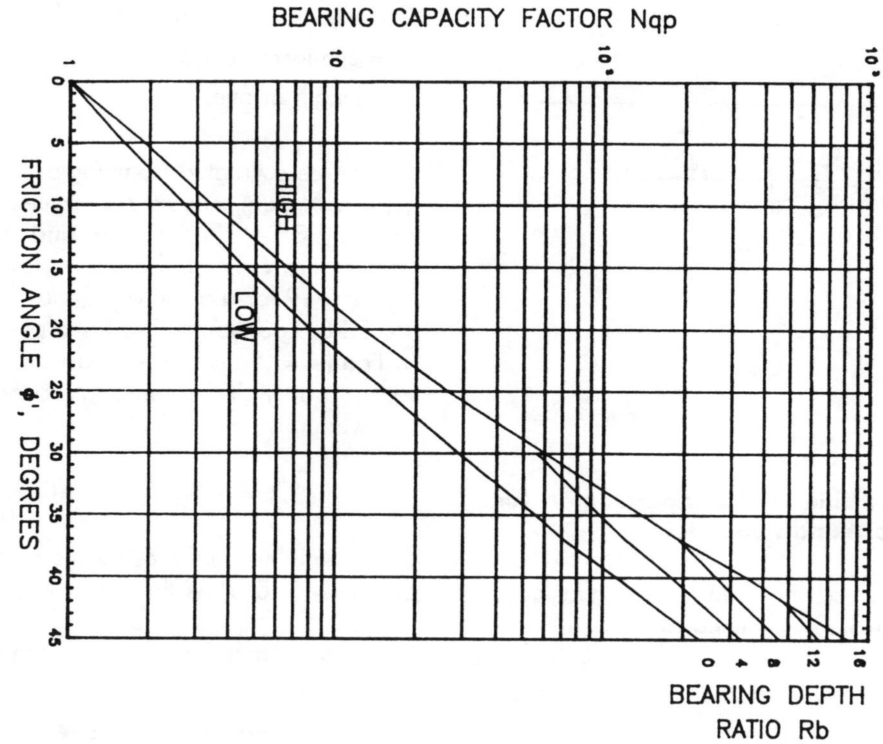

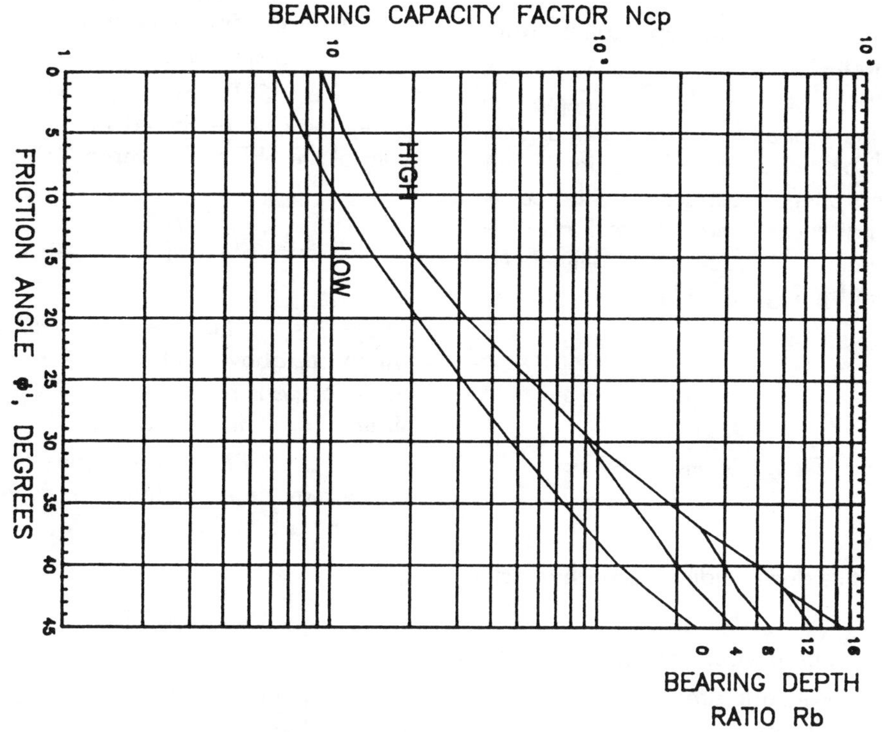

Figure 5-15. Bearing capacity factors for Meyerhof method (Data from Meyerhof 1976)

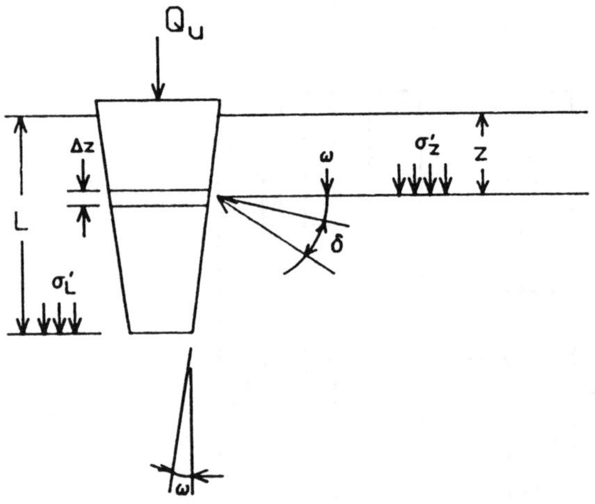

Figure 5-16. Illustration of input parameters for Nordlund's equation

3. If $Rb > Rc$, check to be sure that $q_{bu} \leq q_\ell$ = the limiting stress. The limiting stress is given by

$$q_\ell = N_{qp} \cdot \tan \phi' \qquad (5\text{-}31c)$$

where q_ℓ is in units of ksf.

Refer to Vanikar (1986) for further applications using the Meyerhof method.

b. Nordlund Method. This semi-empirical method considers the shape of the pile taper and the influence of soil displacement on skin friction. Equations for calculating ultimate capacity are based on load test results that include timber, steel *H*, pipe, monotube, and Raymond steptaper piles. Ultimate capacity is determined by (Figure 5-16)

$$Q_u = \alpha_f N_{qp} A_b \sigma'_L + \sum_{z=0}^{z=L} K C_f \sigma'_z \frac{\sin(\delta + \omega)}{\cos \omega} C_z \Delta L \qquad (5\text{-}32a)$$

xnere

α_f = dimensionless pile depth-width relationship factor

A_b = pile point area, ft^2

σ'_L = effective overburden pressure at pile point, ksf

K = coefficient of lateral earth pressure at depth z

C_f = correction factor for K when $\delta \neq \phi'$

ϕ' = effective soil friction angle, degrees

δ = friction angle between pile and soil

ω = angle of pile taper from vertical

σ'_z = effective overburden pressure at the center of depth increment ΔL, $0 < z \leq L$, ksf

C_z = pile perimeter at depth z, ft

ΔL = pile increment, ft

L = length of pile, ft

ϕ' may be estimated from Table 3-1. Point resistance $q_{bu} = \alpha_f N_{qp} \sigma'_L A_p$ should not exceed $q_\ell A_p$ where q_ℓ is given by Equation 5-31c. α_f and N_{qp} may be found from Figure 5-17, K from Figure 5-18, and δ from Figure 5-19 for a given ϕ'. C_f may be found from Figure 5-20. Equation 5-32a for a pile of uniform cross-section (ω = 0) and length L driven in a homogeneous soil with a single friction angle ϕ and single effective unit weight is

$$Q_u = \alpha_f N_{qp} A \sigma'_L + K C_f \sigma'_m \sin\delta C_s L \qquad (5\text{-}32b)$$

where A is the pile cross-section area, C_s is the pile perimeter and σ'_m is the mean effective vertical stress between the ground surface and pile tip, ksf. Table 5-8 provides a procedure for using the Nordlund method (Data from Vanikar 1986).

3. Field Estimates From In Situ Soil Tests. The ultimate end bearing capacity of soils may be estimated from field tests if laboratory soil or other data are not available.

a. SPT Meyerhof Method. End bearing capacity may be estimated from penetration resistance data of the SPT by (Meyerhof 1976)

$$q_{bu} = 0.8 \cdot N_{SPT} \cdot \frac{L_b}{B} < 8 \cdot N_{SPT} \qquad \frac{L}{B} \geq 10 \qquad (5\text{-}33)$$

where N_{SPT} is the average uncorrected blow count within $8B_b$ above and $3B_b$ below the pile tip. L_b is the depth of penetration of the pile tip into the bearing stratum. q_{bu} is in units of ksf.

b. CPT Meyerhof method. End bearing capacity may be estimated from cone penetration resistance data by (Meyerhof 1976)

$$q_{bu} = \frac{q_c}{10} \cdot \frac{L_b}{B} < q_\ell \qquad (5\text{-}34)$$

based on numerous load tests of piles driven to a firm cohesionless stratum not underlain by a weak deposit. q_ℓ is the limiting static point resistance given approximately by Equation 5-31c. N_{qp} should be estimated by the Meyerhof method, Table 4-3. q_{bu} and q_ℓ are in units of ksf.

c. CPT B & G method. End bearing capacity may also be estimated from cone penetration resis-

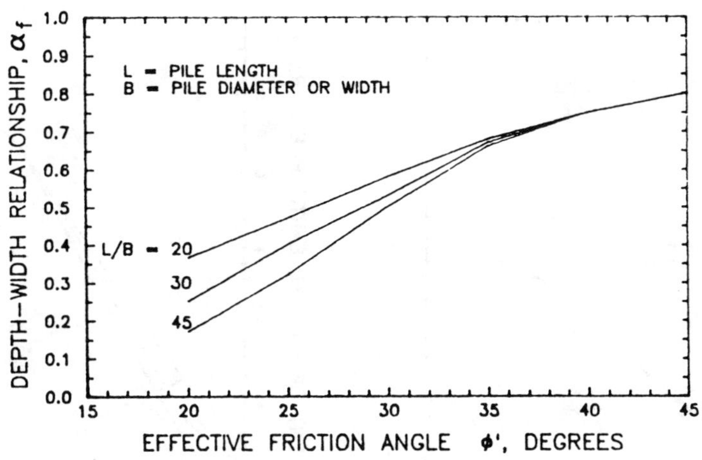

a. α_f

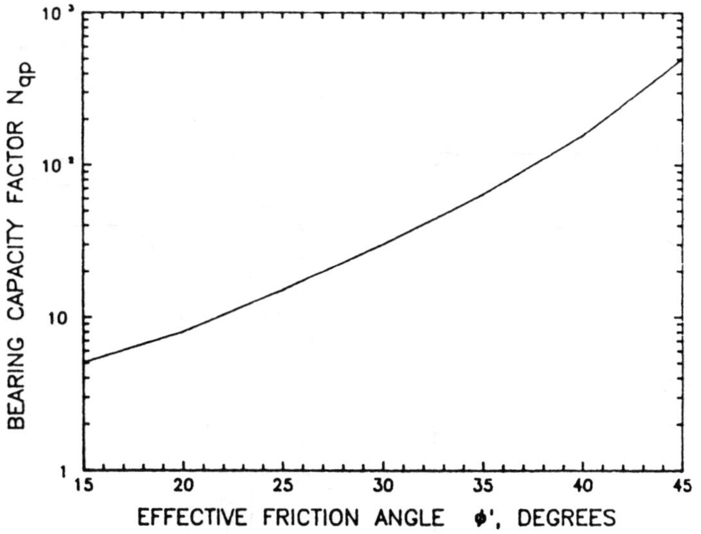

b. N_{qp}

Figure 5-17. Coefficient α_f and bearing capacity factor N_{qp} for the Nordlund method (Data from Vanikar 1986)

tance data by (Bustamante and Gianeselli 1983)

$$q_{bu} = k_c \cdot q_c \qquad (5\text{-}35)$$

where

k_c = point correlation factor, Table 5-9

q_c = average cone point resistance within 1.5*B* below the pile point, ksf

B_b = base diameter, ft

d. CPT 1978 FHWA-Schmertmann method (modified). End bearing capacity may be estimated by (Schmertmann 1978)

$$q_{bu} = \frac{q_{c1} + q_{c2}}{2} \qquad (5\text{-}36)$$

where q_{c1} and q_{c2} are unit cone resistances determined by the procedure described in Figure 5-21. For example, q_{c1} calculated over the minimum path is as

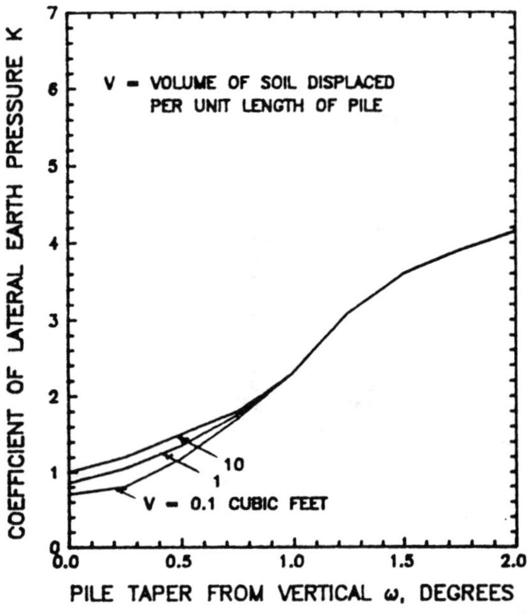

a. $\phi' = 25°$

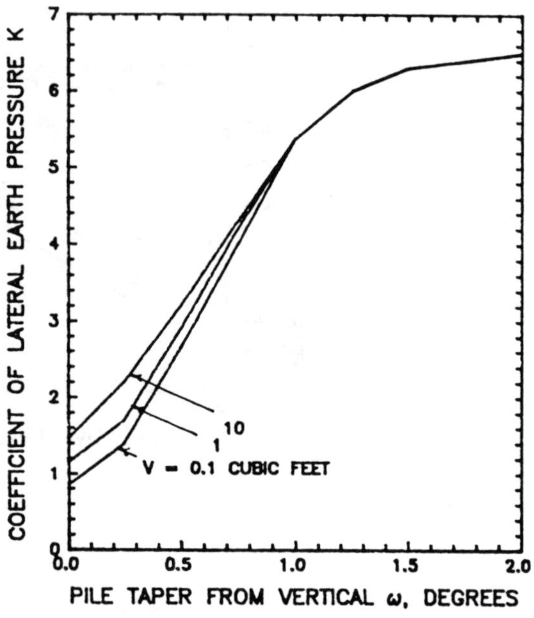

b. $\phi' = 30°$

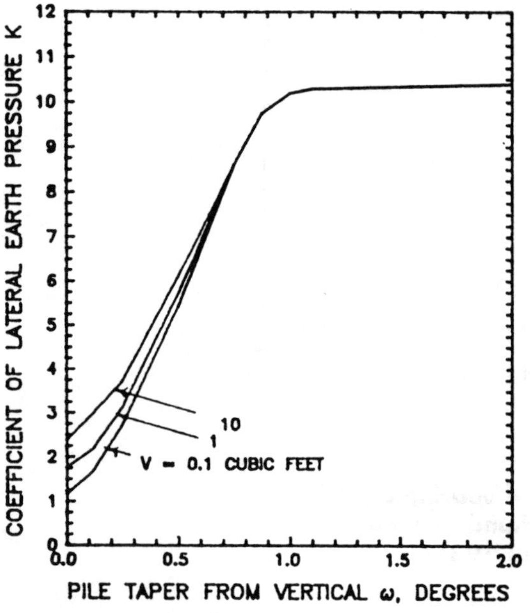

c. $\phi' = 35°$

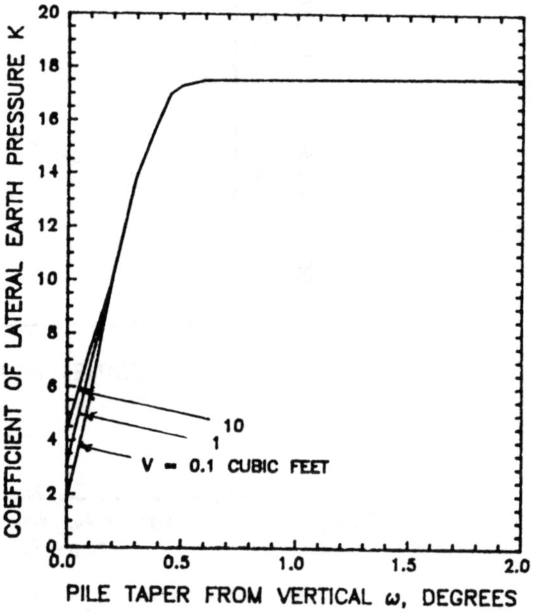

d. $\phi' = 40°$

Figure 5-18. Coefficient *K* for various friction angles ϕ' and pile taper ω for Nordlund method (Data from Vanikar 1986)

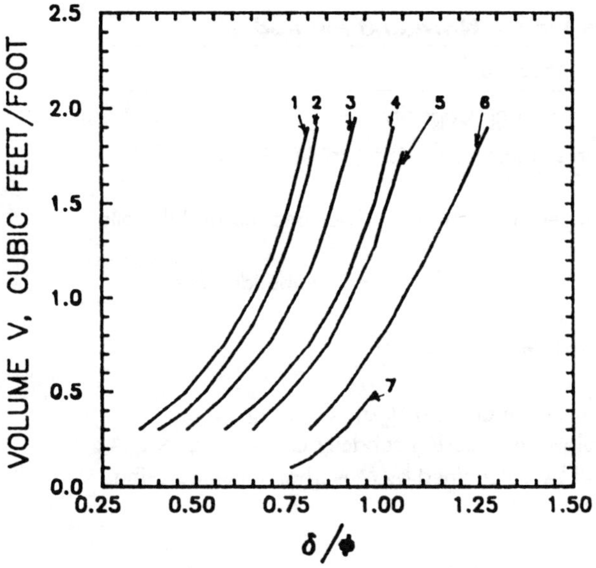

Figure 5-19. Ratio δ/φ for given displacement volume *V* (Data from Vanikar 1986)

follows:

$$q_{c1} = \frac{180 + 170 + 170 + 170 + 170}{5} = 172 \text{ ksf}$$

q_{c2} over the minimum path is as follows:

$$c_{c2}$$

$$= \frac{120 + 150 + 160 + 160 + 160 + 160 + 160 + 160}{8}$$

$$= 153.75 \text{ ksf}$$

From Equation 5-36, $q_{bu} = (172 + 153.75)/2 = 162.9$ ksf.

4. Scale Effects. Ultimate end bearing capacity q_{bu} tends to be less for larger diameter driven piles and drilled shafts than that indicated by Equations 5-33 or 5-34 or Equation 5-2b using Equations 5-31 to estimate N_{cp} or N_{qp} (Meyerhof 1983). Skin friction is independent of scale effects.

a. Sands. The reduction in end bearing capacity has been related with a reduction of the effective angle of internal friction ϕ' with larger diameter deep foundations. End bearing capacity q_{bu} from Equation 5-2 should be multiplied by a reduction factor R_{bs} (Meyerhof 1983)

$$R_{bs} = \left[\frac{B + 1.64}{2B}\right]^m \leq 1 \qquad \text{(5-37a)}$$

for $B > 1.64$ ft. The exponent $m = 1$ for loose sand, 2 for medium dense sand, and 3 for dense sand.

b. Clays. The reduction factor R_{bc} appears related to soil structure and fissures. For driven piles in stiff fissured clay, R_{bc} is given by Equation 5-37a where $m = 1$. For bored piles

$$R_{bc} = \left[\frac{B + 3.3}{2B + 3.3}\right] \leq 1 \qquad \text{(5-37b)}$$

for B from 0 to 5.75 ft.

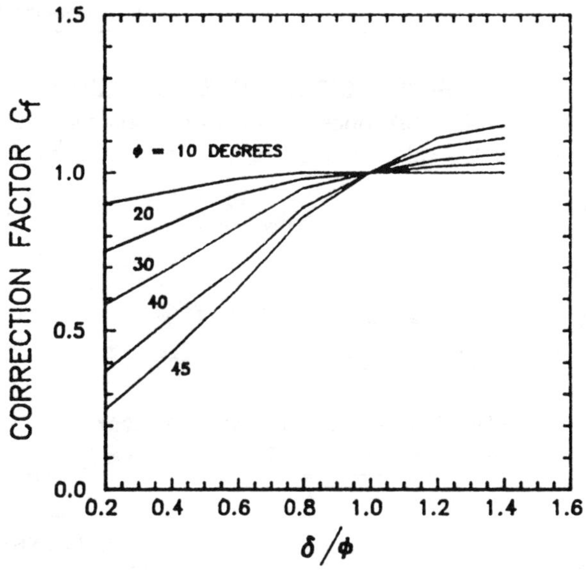

Figure 5-20. Correction factor C_f when $\delta \neq \phi$ (Data from Vanikar 1986)

Table 5-8. Procedure for the Nordlund Method

Step	Procedure
	a. End Bearing Capacity
1.	Determine friction angle ϕ' from in situ test results using Tables 3-1 and 3-2 for each soil layer. $\phi = \phi'$
2.	Determine α_f using ϕ for the soil layer in which the tip is embedded and the pile L/B ratio from Figure 5-17a
3.	Determine N_{qp} using ϕ for the soil layer in which the tip is embedded from Figure 5-17b
4.	Determine effective overburden pressure at the pile tip, σ'_L
5.	Determine the pile point area, A_b
6.	Determine the bearing resistance pressure $q_{bu} = \alpha_f N_{qp}\sigma'_L$. Check $q_{bu} \leq q_\ell = N_{qp}\tan\phi$ of Equation 5-31c. Calculate end bearing capacity $Q_{bu} = q_{bu}A_b \leq q_\ell A_b$. N_{qp} used in Equation 5-31c should be determined by Meyerhof's method using Figure 5-15
	b. Skin Friction Capacity
7.	Compute volume of soil displaced per unit length of pile
8.	Compute coefficient of lateral earth pressure K for ϕ' and ω using Figure 5-18. Use linear interpolation
9.	Determine δ/ϕ for the given pile and volume of displaced soil V from Figure 5-19. Calculate δ for friction angle ϕ
10.	Determine correction factor C_f from Figure 5-20 for ϕ and the δ/ϕ ratio
11.	Calculate the average effective overburden pressure σ'_z of each soil layer
12.	Calculate pile perimeter at center of each soil layer, C_z
13.	Calculate the skin friction capacity of the pile in each soil layer i from $$Q_{sui} = KC_f\sigma'_z \frac{\sin(\delta + \omega)}{\cos\omega} C_z\Delta L$$ Add Q_{sui} of each soil layer to obtain Q_{su}, $Q_{su} = \Sigma\, Q_{sui}$ of each layer
14.	Compute ultimate total capacity, $Q_u = Q_{bu} + Q_{su}$

B. SKIN RESISTANCE CAPACITY. The maximum skin resistance that may be mobilized along an element of pile length ΔL may be estimated by

$$Q_{sui} = A_{si} \cdot f_{si} \tag{5-9}$$

where

A_{si} = area of pile element i, $C_{si} \cdot \Delta L$, ft^2

C_{si} = shaft circumference at pile element i, ft

Table 5.9 Point Correlation Factor k_c (Bustamante and Gianeselli 1983)

	k_c	
Soil	Driven Pile	Drilled Shaft
Clay–Silt	0.600	0.375
Sand–Gravel	0.375	0.150
Chalk	0.400	0.200

L = length of pile element, ft

f_{si} = skin friction at pile element i, ksf

1. Cohesive Soil.

a. Alpha method. The skin friction of a length of pile element may be estimated by

$$f_{si} = \alpha_a \cdot C_u \tag{5-10}$$

where

α_a = adhesion factor

C_u = undrained shear strength, ksf

Local experience with existing soils and load test results should be used to estimate appropriate α_a. Estimates of α_a may be made from Table 5-10 in the absence of load test data and for preliminary design.

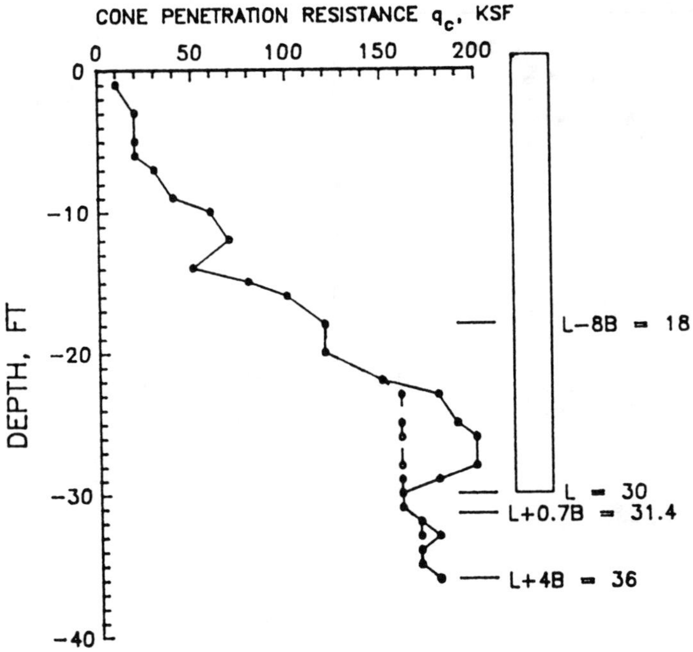

$$q_{bu} = \frac{q_{c1} + q_{c2}}{2}$$

q_{c1} = Average q_c over a distance of L+0.7B to L+4B below the pil tip. Use the minimum path. minimum path is a vertical line that spans the minimum measured q_c between L+0.7B and L+4B.

q_{c2} = Average q_c over a distance of L to L-8B above the pile tip. Use the minimum path as above.

Figure 5-21. Estimating pile tip capacity from CPT data (Data from Schmertmann 1978)

b. Lambda Method. This semi-empirical method is based on numerous load test data of driven pipe piles embedded in clay assuming that end bearing capacity was evaluated from Equation 5-2a using $N_{cp} = 9$ and $\zeta_{cp} = 1$ (Vijayvergiya and Focht 1972). The N_{qp} and $N_{\gamma p}$ terms are not used. Skin friction is

$$f_{si} = \lambda \cdot (\sigma'_m + 2C_{um}) \qquad \text{(5-38a)}$$

where

λ = correlation factor (Figure 5-22)

σ'_m = mean effective vertical stress between the ground surface and pile tip, ksf

C_{um} = mean undrained shear strength along the pile length, ksf

Table 5-10. Adhesion Factors for Driven Piles in Cohesive Soil (Data from Tomlinson 1980)

Length/Width Ratio $\frac{L}{B}$	Undrained Shear Strength, C_u, ksf	Adhesion Factor α_a
< 20	< 3	$1.2-0.3C_u$
	> 3	0.25
> 20	0.0–1.5	1.0
	1.5–4.0	$1.5-0.4C_u$
	> 4	0.3

λ may also be given approximately by

$$\lambda = L^{-0.42} \qquad L \geq 10 \text{ FT} \qquad \text{(5-38b)}$$

where L is the pile length, ft.

2. Cohesionless Soil. The soil-shaft skin friction may be estimated using effective stresses

$$f_{si} = \beta_f \cdot \sigma'_i \qquad \text{(5-12a)}$$

$$\beta_f = K \cdot \tan \delta_a \qquad \text{(5-12b)}$$

where

β_f = lateral earth pressure and friction angle factor

K = lateral earth pressure coefficient

δ_a = soil-shaft effective friction angle, $\leq \phi'$, deg

σ'_i = effective vertical stress in soil in pile element i, ksf

Cohesion c is taken as zero.

a. Figure 5-5 indicates appropriate values of β_f as a function of the effective friction angle ϕ' of the soil prior to installation of the deep foundation.

b. The effective vertical stress σ'_i approaches a limiting stress at the critical depth Lc, then remains constant below Lc. Lc may be estimated from Figure 5-3.

c. The Nordlund method in Table 5-8b provides an alternative method of estimating skin resistance.

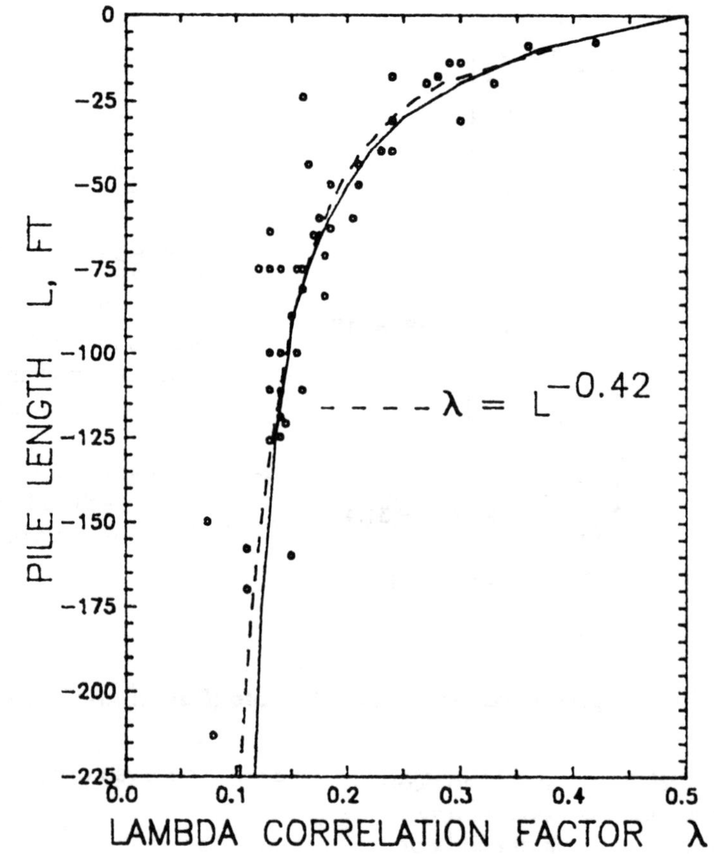

Figure 5-22. Lambda correlation factor (Data from Vijayvergiya and Focht 1972)

3. CPT Field Estimate. The skin friction f_{si} may be estimated from the measured cone resistance q_c for the piles described in Table 5-2b using the curves given in Figure 5-6 for clay and silt, sand and gravel, and chalk (Bustamante and Gianeselli 1983).

C. ULTIMATE CAPACITY FROM WAVE EQUATION ANALYSIS. Estimates of total bearing capacity may be performed using computer program GRLWEAP (Goble Rausche Likins and Associates, Inc. 1988). The analysis uses wave propagation theory to calculate the force pulse transmitted along the longitudinal pile axis caused by impact of the ram (Figure 5-23). The force pulse travels at a constant velocity depending on the pile material and this pulse is attenuated by the soil frictional resistance along the embedded length of the pile. The pile penetrates into the soil when the force pulse reaching the pile tip exceeds the ultimate soil resistance at the pile tip Q_{ub}. Program GRLWEAP and user's manual are licensed to the Waterways Experiment Station and it is available to the US Army Corps of Engineers.

1. Description. The pile driving and soil system consists of a series of elements supported by linear elastic springs and dashpots which have assumed parameters (Figure 5-23). Characteristics of commonly used pile hammers and piles are available in the program data files driving systems. Input parameters include the dynamic damping constants for each dashpot, usually in units of seconds/inch, ultimate soil resistance Q_u in kips and quake in fractions of an inch for each spring. Each dashpot and spring represent a soil element. The quake of the pile is its displacement at Q_{ub}. Input data for Q_{ub}, quake, and ultimate skin resistance of each element Q_{sui} are usually assumed. Actual load distribution data are normally not available and require results of instrumented load tests. Standard values are available in the user's manual for soil input parameters.

2. Analysis. The wave equation analysis provides a relationship between the pile capacity and the driving resistance in blows per inch (or blows per foot if needed). This relationship can be developed for different pile lengths and then used in the field when the pile has been driven sufficiently to develop the required capacity. The wave equation can also be used

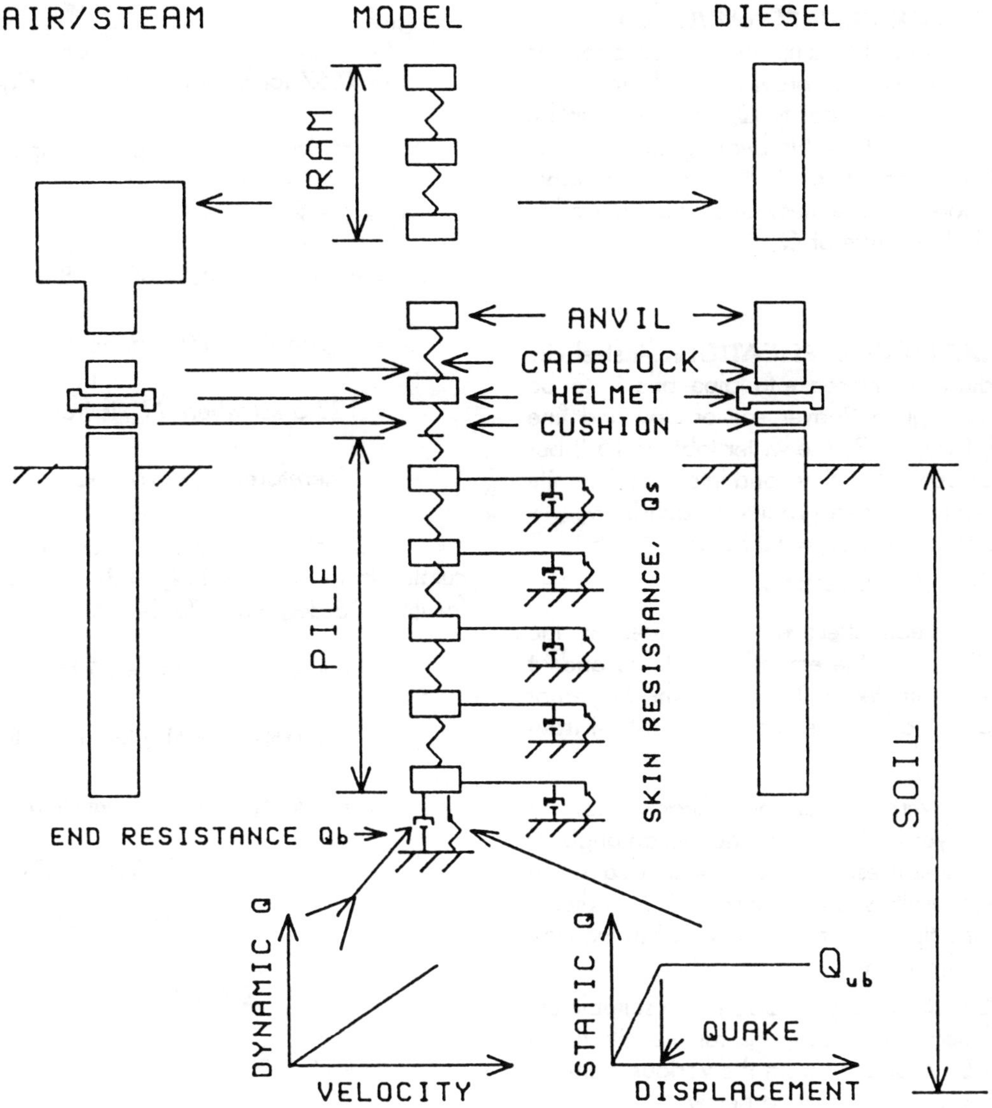

Figure 5-23. Schematic of wave equation model

to develop relationships between driving stresses in the pile and penetration resistance for different combinations of piles and pile driving equipment.

3. Application. The wave equation analysis is used to select the most suitable driving equipment to ensure that the piles can develop the required capacity and select the minimum pile section required to prevent overstressing the pile during driving.

4. Calibration. Calculations from program GRLWEAP may be calibrated with results of dynamic load tests using pile driving analyzer (PDA) equipment. The force and velocity versus time curves calculated from GRLWEAP are adjusted to agree with the force and velocity versus time curves measured by the PDA during pile driving or during a high strain test. A high strain causes a force at the pile tip sufficient to exceed the ultimate soil resistance Q_{ub}. Drilled shafts may be analyzed with the PDA during a high strain test where heavy loads are dropped by a crane on the head of the shaft.

5. Factors of Safety. In general, pile capacity calculated by GRLWEAP should be divided by a factor of safety FS = 3 to estimate allowable capacity or FS = 2.5 if calibrated with results of dynamic load tests. If load tests are performed, FS = 2 can be used with GRLWEAP.

6. Restriking. Soils subject to freeze or relaxation could invalidate a wave equation analysis; therefore, installed piles should be tested by restriking while using PDA equipment after a minimum waiting period following installation such as 1 day or more as given in the specifications.

D. PILE DRIVING FORMULAS. Pile driving formulas, Table 5-11, although not as good as wave equation analysis, can provide useful, simple estimates of ultimate pile capacity Q_u and they can be obtained quickly. The allowable bearing capacity can be estimated from Equations 1-2 using FS in Table 5-11. Two or more of these methods should be used to provide a probable range of Q_u.

E. EXAMPLE APPLICATION. A steel circular 1.5-ft diameter closed end pipe pile is to be driven 30 ft through a 2-layer soil of clay and fine uniform sand, Figure 5-7. The water table is 15 ft below ground surface at the clay-sand interface. The pile will be filled with concrete grout with density γ_{conc} = 150 lbs/ft^3. Design load Q_d = 100 kips.

1. Soil Parameters.

a. The mean effective vertical stress in the sand layer adjacent to the embedded pile σ'_s and at the pile tip σ'_L are limited to 1.8 ksf for the Meyerhof and Nordlund methods. Otherwise, σ'_L is 2.4 ksf from Equation 5-13b.

b. The average undrained shear strength of the upper clay layer is C_u = 2 ksf. The friction angle of the lower sand layer is estimated at ϕ' = 36 deg. Cone penetration test results shown in Figure 5-21 indicate an average cone tip resistance q_c = 40 ksf in the clay and 160 ksf in the sand.

2. End Bearing Capacity. A suitable estimate of end bearing capacity q_{bu} for the pile tip in the sand may be evaluated from the various methods for cohesionless soil as described below.

a. Meyerhof Method. From Figure 5-3, Rc = 10 and $Lc = Rc \cdot B = 10 \cdot 1.5$ or 15 ft for ϕ' = 36 deg. N_{qp} from Figure 5-15 is 170 for $R_b = L/B = 15/1.5 = 10$. From Equation 5-2c with limiting pressure q_ℓ from Equation 5-31c

$$q_{bu} = \sigma'_L \cdot N_{qp} \leq N_{qp} \cdot \tan\phi' \text{ if } L > Lc$$

$$\sigma'_L \cdot N_{qp} = 1.8 \cdot 170 = 306 \text{ ksf}$$

$$N_{qp} \cdot \tan\phi' = 170 \cdot \tan 36 = 123.5 \text{ ksf}$$

$$q_{bu} \leq q_\ell, \text{ therefore } q_{bu} = 123.5 \text{ ksf}$$

b. Nordlund Method. The procedure in Table 5-8a may be used to estimate end bearing capacity.

$$\alpha_f = 0.67 \text{ for } \phi' = 36 \text{ deg from Figure 5-17a}$$

$$N_{qp} = 80 \text{ for } \phi' = 36 \text{ deg from Figure 5-17b}$$

$$\sigma'_L = 1.8 \text{ ksf}$$

$$q_{up} = \alpha_f N_{qp} \sigma'_L = 0.67 \cdot 80 \cdot 1.8 = 96.5 \text{ ksf}$$

$$q_\ell = N_{qp} \tan\phi' = 170 \cdot \tan 36$$

$$= 123.5 \text{ ksf where } N_{qp} \text{ is from Figure 5-15.}$$

Therefore, $q_{bu} = 96.5 \text{ ksf} \leq q_\ell$

c. Hansen Method. From Table 4-5 (or calculated from Table 4-4) N_{qp} = 37.75 and $N_{\gamma p}$ = 40.05 for ϕ' = 36 deg. From Table 4-5,

$$\zeta_{qs} = 1 + \tan\phi = 1 + \tan 36 = 1.727$$

$$\zeta_{qd} = 1 + 2\tan\phi(1 - \sin\phi)^2 \cdot \tan^{-1}(L_{sand}/B)$$

$$= 1 + 2\tan 36(1 - \sin 36)^2 \cdot \tan^{-1}(15/1.5) \cdot \pi/180$$

$$= 1 + 2 \cdot 0.727(1 - 0.588)^2 \cdot 1.471 = 1.363$$

$$\zeta_{qp} = \zeta_{qs} \cdot \zeta_{qd} = 1.727 \cdot 1.363 = 2.354$$

$$\zeta_{\gamma s} = 1 - 0.4 = 0.6$$

$$\zeta_{\gamma d} = 1.00$$

$$\zeta_{\gamma p} = \zeta_{\gamma s} \cdot \zeta_{\gamma d} = 0.6 \cdot 1.00 = 0.6$$

From Equation 5-2a

$$q_{bu} = \sigma'_L \cdot N_L \cdot \zeta_{qp} + (B_b/2) \cdot \gamma'_s \cdot N_{\gamma p} \cdot \zeta_{\gamma p}$$

$$= 2.4 \cdot 37.75 \cdot 2.354 + (1.5/2) \cdot 0.04 \cdot 40.05 \cdot 0.6$$
$$= 213.3 + 0.7 = 214 \text{ ksf}$$

The $N_{\gamma p}$ term is negligible and could have been omitted.

d. Vesic Method. The reduced rigidity index from Equation 5-5c is

$$I_{rr} = \frac{I_r}{1 + \epsilon_v \cdot I_r}$$

$$= \frac{57.3}{1 + 0.006 \cdot 57.3} = 42.6$$

Table 5-11. Pile Driving Formulas

Method	Equation for Ultimate Bearing Capacity Q_u, kips	Factor of Safety
Gates	$27(E_h E_r)^{1/2}(1 - \log_{10}S)$	3
Pacific Coast Uniform Building Code	$\frac{12E_hE_rC_{p1}}{S + C_{p2}}$, $C_{p1} = \frac{W_r + c_pW_p}{W_r + W_p}$, $C_{p2} = \frac{Q_uL}{AE_p}$ c_p = 0.25 for steel piles; = 0.10 for other piles Initially assume $C_{p2} = 0$ and compute Q_u; reduce Q_u by 25 percent, compute C_{p2}, then recompute Q_u; compute a new C_{p2}, compute Q_u until Q_u used = Q_u computed	4
Danish	$\frac{12E_hE_r}{S + C_d'}$, $C_d = \left[\frac{144E_hE_rL}{2AE_p}\right]^{1/2}$, inches	3–6
Engineering News Record	Drop Hammers: $\frac{12W_rh}{S + 1.0}$	6
	Other Hammes: $\frac{24E_r}{S + 0.1}$	6

Nomenclature:
- A = area of pile cross-section, ft²
- E_h = hammer efficiency
- E_p = pile modulus of elasticity, ksf
- E_r = manufacturer's hammer-energy rating (or W_rh), kips-ft
- h = height of hammer fall, ft
- L = pile length, inches
- S = average penetration in inches per blow for the last 5 to 10 blows for drop hammers and 10 to 20 blows for other hammers
- W_r = weight of striking parts of ram, kips
- W_p = weight of pile including pile cap, driving shoe, capblock and anvil for double-acting steam hammers, kips

where

(Equation 5-5e):

$$\epsilon_v = \frac{1 - 2 \cdot \mu_s}{2(1 - \mu_s)} \cdot \frac{\sigma_L'}{G_s}$$

$$= \frac{1 - 2 \cdot 0.3}{2(1 - 0.3)} \cdot \frac{2.4}{100} = 0.006$$

(Equation 5-5d):

$$I_r = \frac{G_s}{\sigma_L' \cdot \tan\phi'} = \frac{100}{2.4 \cdot \tan 36} = 57.3$$

From Equation 5-5b

$$N_{qp} = \frac{3}{3 - \sin\phi'} \cdot e^{\frac{(90-\phi')\pi}{180}\tan\phi'}$$

$$\cdot \tan^2\left[45 + \frac{\phi'}{2}\right] \cdot I_{rr}^{\frac{4\sin\phi'}{3(1+\sin\phi')}}$$

$$N_{qp} = \frac{3}{3 - \sin 36}$$

$$\cdot e^{\frac{(90-36)\pi}{180}\tan 36} \cdot \tan^2\left[45 + \frac{36}{2}\right] \cdot I_{rr}^{\frac{4\sin 36}{3(1+\sin 36)}}$$

$$N_{qp} = \frac{3}{3 - 0.588} e^{0.685} \cdot 3.852 \cdot 42.6^{0.494}$$

$$N_{qp} = 1.244 \cdot 1.984 \cdot 3.852 \cdot 6.382 = 60.7$$

The shape factor from Equation 5-6a is

$$\zeta_{qp} = \frac{1 + 2K_o}{3} = \frac{1 + 2 \cdot 0.42}{3} = 0.61$$

where K_o was evaluated using Equation 5-6c. From Equation 5-2c,

$$q_{bu} = \sigma_L' \cdot N_{qp} \cdot \zeta_{qp}$$

$$= 2.4 \cdot 60.7 \cdot 0.61 = 88.9 \text{ ksf}$$

e. General Shear Method. From Equation 5-8

$$N_{qp} = \frac{e^{\frac{270-\phi'}{180}\cdot \tan\phi'}}{2\cos^2\left[45 + \frac{\phi'}{2}\right]} = \frac{e^{\frac{270-36}{180}\cdot \pi\tan 36}}{2\cos^2\left[45 + \frac{\phi'}{2}\right]}$$

$$= \frac{e^{1.3\pi \cdot 0.727}}{2 \cdot 0.206}$$

$$N_{qp} = \frac{19.475}{0.412} = 47.24$$

The shape factor $\zeta_{qp} = 1.00$ when using Equation 5-8. From Equation 52c,

$$q_{bu} = \sigma'_L \cdot N_{qp} \cdot \zeta_{qp}$$

$$= 2.4 \cdot 47.24 \cdot 1.00 = 113.4 \text{ ksf}$$

f. CPT Meyerhof Method. From Equation 5-34

$$q_{bu} = \frac{q_c}{10} \cdot \frac{L_{sand}}{B} < q_\ell$$

where $q_\ell = N_{qp} \cdot \tan\phi'$ ksf. Substituting the parameters into Equation 5-34

$$q_{bu} = \frac{160}{10} \cdot \frac{15}{1.5} = 160 \text{ ksf}$$

The limiting q_ℓ is 123.5 ksf, therefore q_{bu} = 123.5 ksf

g. CPT B & G. From Equation 5-35

$$q_{bu} = k_c \cdot q_c$$

where k_c = 0.375 from Table 5-9. q_{bu} = 0.375 · 160 = 60 ksf.

h. CPT FHWA & Schmertmann. The data in Figure 5-21 can be used with this method to give q_{bu} = 162.9 ksf as in the example illustrating this method in paragraph 5-7a.

i. Comparison of Methods. A comparison of methods is shown as follows:

Method	q_{bu}, ksf
Meyerhof	124
Nordlund	97
Hansen	214
Vesic	89
General Shear	113
CPT Meyerhof	124
CPT B & G	60
CPT FHWA & Schmertmann	163

These calculations indicate q_{bu} from 60 to 214 ksf. Discarding the highest (Hansen) and lowest (CPT B & G) values gives an average q_{bu} = 118 ksf. Scale effects of Equations 5-37 are not significant because B < 1.64 ft.

2. Skin Friction Capacity. A suitable estimate of skin friction f_s from the soil-shaft interface may be evaluated for both the clay and sand as illustrated below.

a. Cohesive Soil. The average skin friction using the Alpha method from Equation 5-10 is

$$f_s = \alpha_a \cdot C_u = 0.6 \cdot 2 = 1.2 \text{ ksf}$$

where $\alpha_a = 1.2 - 0.3C_u = 0.6$ from Table 5-11 and $L/B < 20$. The average skin friction using the Lambda method from Equation 5-38a and using L = 15 ft for the penetration of the pile only in the clay is

$$f_s = \lambda(\sigma'_m + 2C_{um}) = 0.32(0.9 + 2 \cdot 2) = 1.57 \text{ ksf}$$

where

$$\lambda = L^{-0.42} = 15^{-0.42} = 0.32 \text{ from Equation 5-38b}$$

$$\sigma'_m = \frac{D_c}{2} \cdot \gamma'_{clay} = \frac{15}{2} \cdot 0.12 = 0.9 \text{ ksf}$$

A reasonable average value of skin friction is 1.4 ksf for the clay.

b. Cohesionless Soil. The average skin friction from Equation 5-12a using σ'_s limited to 1.8 ksf is

$$f_s = \beta_f \cdot \sigma'_s = 0.96 \cdot 1.8 = 1.7 \text{ ksf}$$

where β_f is found from Figure 5-5 using ϕ' = 36 deg. The Nordlund method of Table 5-8b provides an alternative estimate

$$V = \pi \cdot (1.5^2/2) \cdot 1 = 1.77 \text{ ft}^3/\text{ft}$$

$K = 2.1$ from Figure 5-18 for $\omega = 0$ deg

$\delta/\phi = 0.78$ for $V = 1.77$ and pile type 1 from Figure 5-19

$$\delta = 0.78 \cdot 36 = 28 \text{ deg}$$

$C_f = 0.91$ for $\delta/\phi = 0.78$ and $\phi = 36$ deg from Figure 5-20

$\sigma'_z = 1.8$ ksf limiting stress

$$C_z = \pi \cdot B_s = \pi \cdot 1.5 = 4.71 \text{ ft}$$

$$Q_{sz} = KC_f\sigma'_z\sin\delta \cdot C_z\Delta z = 2.1 \cdot 0.91 \cdot 1.8 \cdot \sin 28 \cdot 4.71 \cdot 15 = 114 \text{ kips}$$

$$f_s = Q_{sz}/(C_z\Delta z) = 1.6 \text{ ksf}$$

Skin friction for the sand is about 1.6 ksf.

c. CPT Field Estimate. The driven pile is described as "steel" from Table 5-2b. Curve 1 of Clay and Silt, Figure 5-6a, and curve 1 of Sand and Gravel, Figure 5-6b, should be used. From these figures, f_s of the clay is 0.7 ksf and f_s of the sand is 0.7 ksf.

d. Comparison of Methods. Skin friction varies from 0.7 to 1.4 ksf for the clay and 0.7 to 1.6 ksf for the sand. Skin friction is taken as 1.0 ksf in the clay and 1 ksf in the sand.

3. Ultimate Total Capacity. The total bearing capacity from Equation 5-1a is

$$Q_u = Q_{bu} + Q_{su} - W_p$$

where

$$W_p = \frac{\pi B^2}{4} \cdot L \cdot \gamma_{conc}$$

$$= \frac{\pi \cdot 1.5^2}{4} \cdot 30 \cdot \frac{150}{1000}$$

$$= 8 \text{ kips for the pile weight}$$

a. Q_{bu} may be found from Equation 5-1b

$$Q_{bu} = A_b \cdot q_{bu} = 1.77 \cdot 118 = 209 \text{ kips}$$

where A_b = area of the base, $\pi B^2/4 = \pi \cdot 1.5^2/4 = 1.77$ ft^2.

b. Q_{su} may be found from Equation 5-1b and 5-9

$$Q_{su} = \sum_{i=1}^{n} Q_{sui} = C_s \cdot \Delta L \sum_{i=1}^{2} f_{si}$$

where $C_s = \pi B$. Therefore,

$$Q_{su} = \pi B \cdot (L_{sand} \overset{sand}{f_s} + L_{clay} \overset{clay}{f_s})$$

$$= \pi \cdot 1.5 \cdot 15(1.0 + 1.0) = 141 \text{ kips}$$

c. Inserting end bearing and skin resistance bearing capacity values into Equation 5-1a,

$$Q_u = 209 + 141 - 8 = 342 \text{ kips}$$

The minimum and maximum values of q_{bu} and f_s calculated above could be used to obtain a range of Q_u if desired.

4. Allowable Bearing Capacity. The allowable bearing capacity from Equation 1-2b using $FS = 3$ is

$$Q_a = \frac{Q_s}{FS} = \frac{337}{3} = 112 \text{ kips}$$

5-8. Lateral Load Capacity of Single Piles

Evaluation of lateral load capacity is treated similarly to that for single drilled shafts in 5-4. Lateral load capacity may be determined by load tests, by analytical methods such as Broms' equations or *p-y* curves and by arbitrary values. Most piles are placed in groups where group capacity controls performance.

A. LOAD TESTS. Lateral load tests are economically justified for large projects and may be performed as described in ASTM D 3966.

B. ANALYTICAL METHODS.

1. Program COM624G using *p-y* curves are recommended for complicated soil conditions.

Table 5-12. Recommendations for Allowable Lateral Pile Loads (Data from Vanikar 1986)

Pile	Allowable Deflection, in.	Allowable Lateral Load, kips			Reference
Timber			10		New York State Department of Transportation 1977
Concrete			15		
Steel			20		
All	0.375		2		New York City Building Code 1968
All	0.25		1 (soft clays)		Teng
Timber	0.25		9		Feagin
Timber	0.50		14		Feagin
Concrete	0.25		12		Feagin
Concrete	0.50		17		Feagin
		Medium sand	Fine sand	Medium clay	McNulty
12 inch Timber (free)	0.25	1.5	1.5	1.5	
12 inch Timber (fixed)	0.25	5.0	4.5	4.0	
16 inch Concrete	0.25	7.0	5.5	5.0	

2. Broms' equations in Table 5-5 can give useful estimates of ultimate lateral loads for many cases.

C. ARBITRARY VALUES.

1. Table 5-12 provides allowable lateral loads for piles.

2. Piles can sustain transient horizontal loads up to 10 percent of the allowable vertical load without considering design features.

5-9. Capacity of Pile Groups

Driven piles are normally placed in groups with spacings less than 8 times the pile diameter or width $8B_s$ and joined at the ground surface by a concrete slab referred to as a pile cap. The capacity of the pile group can be greater than the sum of the capacities of the individual piles because driving compacts the soil and can increase skin friction and end bearing capacity. FS for pile groups should be 3.

A. AXIAL CAPACITY. Deep foundations where spacings between individual piles are less than $8B_s$ cause interaction effects between adjacent piles from overlapping of stress zones in the soil, Figure 5-13. In situ soil stresses from pile loads are applied over a much larger area leading to greater settlement and bearing failure at lower total loads.

1. Optimum Spacing. Piles in a group should be spaced so that the bearing capacity of the group sum of the individual piles. Pile spacings should not be less than $2.5B_s$. The optimum pile spacing is 3 to $3.5B_s$ (Vesic 1977) or greater than $0.02L + 2.5B_s$ where L is the pile length in feet (Canadian Geotechnical Society 1985).

2. Cohesive Soils. Group capacity may be estimated by efficiency and equivalent methods similar for drilled shafts as described in paragraph 5-5a.

3. Cohesionless Soil. Group capacity should be taken as the sum of the individual piles.

B. LATERAL LOAD CAPACITY. Response of pile groups to lateral load requires lateral and axial load soil-structure interaction analysis with assistance of a finite element computer program.

1. Widely Spaced Piles. Where piles are spaced > $7B_s$ or far enough apart that stress transfer is minimal and loading is by shear, the ultimate lateral load of the group T_{ug} is the sum of individual piles. The capacity of each pile may be estimated by methodology in 5-4.

2. Closely Spaced Piles. The solution of ultimate lateral load capacity of closely spaced pile groups require analysis of a nonlinear soil-pile system. Refer to Poulos (1971a), Poulos (1971b), and Reese (1986) for detailed solution of the lateral load capacity of each pile by the Poulos-Focht-Koch method.

3. Group Behavior as a Single Pile. A pile group may be simulated as a single pile with diameter C_g/π where C_g is the pile circumference given as the minimum length of a line that can enclose the group of piles. The moment of inertia of the pile group is $n \cdot I_p$ where I_p is the moment of inertia of a single pile. Program COM624G may be used to evaluate lateral load-deflection behavior of the simulated single pile for given soil conditions. A comparison of results between the Poulos-Focht-Koch and simulated single pile methods was found to be good (Reese 1986).

C. COMPUTER ASSISTED ANALYSIS.

Computer programs are available from the Waterways Experiment Station to assist in analysis and design of pile groups. Refer to EM 1110-2-2906 for further guidance on the analysis of pile groups.

1. Program CPGA. Pile Group Analysis computer program CPGA is a stiffness analysis in three-dimensions assuming linear elastic pile-soil interaction and a rigid pile cap (Hartman et al 1989). Program CPGA uses matrix methods to incorporate position and batter of piles and piles of different sizes and materials. Computer program CPGG displays the geometry and results of program CPGA (Jaeger, Jobst, and Martin 1988).

2. Program CPGC. Pile Group Concrete computer program CPGC develops the interaction diagrams and data required to investigate the structural capacity of prestressed concrete piles (Strom, Abraham, and Jones 1990).

APPENDIX A

REFERENCES

1. TM 5-800-08, Engineering Use of Geotextiles.
2. TM 5-818-1, Soil and Geology Procedures for Foundation Design of Buildings and Other Structures (Except Hydraulic Structures).
3. TM 5-818-5, Dewatering and Groundwater Control.
4. TM 5-818-7, Foundations in Expansive Soils.
5. TM 5-852-4, Arctic and Subarctic Construction Foundations for Structures.
6. TM 5-852-6, Arctic and Subarctic: Construction Calculation Methods for Determination of Depths of Freeze and Thaw in Soils.
7. EM 1110-1-1804, Geotechnical Investigations.
8. EM 1110-2-1902, Stability of Earth and Rockfill Dams.
9. EM 1110-1-1904, Settlement Analysis.
10. EM 1110-2-1906, Laboratory Soils Testing.
11. EM 1110-2-1907, Soil Sampling.
12. EM 1110-2-1913, Design and Construction of Levees.
13. EM 1110-2-2906, Design of Pile Structure and Foundations.
14. EM 1110-2-3506, Grouting Technology.
15. Department of the Navy, 1982. "Foundations and Earth Structures," Report No. NAVFAC DM 7.2. Available from Naval Facilities Engineering Command, 200 Stovall Street, Alexandria, VA 22332.
16. Department of the Navy, 1983. "Soil Dynamics, Deep Stabilization and Special Geotechnical Construction," Report No. NAVFAC DM 7.3. Available from Naval Facilities Engineering Command, 200 Stovall Street, Alexandria, VA 22332.
17. American Concrete Institute (ACI) Committee 318, 1980. "Use of Concrete in Buildings: Design, Specifications and Related Topics," Parts 3 and 4, ACI Manual of Concrete Practice, Available from American Concrete Institute, P. O. 19150, Redford Station, Detroit, MI 48219.
18. American Society for Testing and Materials Standard D 653, "Terminology Relating to Soil, Rock, and Contained Fluids," Vol 04.08. Available from American Society for Testing and Materials, 1916 Race Street, Philadelphia, PA 19103.
19. American Society for Testing and Materials Standard D 854, "Standard Test Method for Specific Gravity of Soils," Vol 04.08. Available from American Society for Testing and Materials, 1916 Race Street, Philadelphia, PA 19103.
20. American Society for Testing and Materials Standard Test Method D 1143, "Piles Under Static Axial Compressive Load," Vol 04.08. Available from American Society for Testing and Materials, 1916 Race Street, Philadelphia, PA 19103.
21. American Society for Testing and Materials Standard Test Method D 1194, "Bearing Capacity of Soil for Static Load and Spread Foodings," Vol 04.08. Available from American Society for Testing and Materials, 1916 Race Street, Philadelphia, PA 19103.
22. American Society for Testing and Materials Standard Test Method D 1556, "Density of Soil in Place by the Sand-Cone Method," Vol 04.08. Available from American Society for Testing and Materials, 1916 Race Street., Philadelphia, PA 19103.
23. American Society for Testing and Materials Standard Test Method D 1586, "Penetration Test and Split-Barrel Sampling of Soils," Vol 04.08. Available from American Society for Testing and Materials, 1916 Race Street, Philadelphia, PA 19103.
24. American Society for Testing and Materials Standard Test Method D 2573, "Field Vane Shear Test in Cohesive Soil," Vol 04.08. Available from American Society for Testing and Materials, 1916 Race Street, Philadelphia, PA 19103.
25. American Society for Testing and Materials Standard Test Method D 2850, "Unconsolidated, Undrained Compressive Strength of Cohesive Soils," Vol 04.08. Available from American Society for Testing and Materials, 1916 Race Street, Philadelphia, PA 19103.
26. American Society for Testing and Materials Standard Test Method D 3441, "Deep, Quasi-Static, Cone and Friction-Cone Penetration Tests of Soils," Vol 04.08. Available from American Society for Testing and Materials, 1916 Race Street, Philadelphia, PA 19103.
27. American Society for Testing and Materials Standard Test Method D 3966, "Piles Under Lateral Loads," Vol 04.08. Available from American So-

ciety for Testing and Materials, 1916 Race Street, Philadelphia, PA 19103.

28. American Society for Testing and Materials Standard Test Methods D 4254, "Minimum Index Density of Soils and Calculation of Relative Density," Vol 04.08. Available from American Society for Testing and Materials, 1916 Race Street, Philadelphia, PA 19103.
29. American Society for Testing and Materials Standard Test Methods D 4546, "One-Dimensional Swell or Settlement Potential of Cohesive Soils," Vol 04.08. Available from American Society for Testing and Materials, 1916 Race Street, Philadelphia, PA 19103.
30. American Society for Testing and Materials Standard Test Method D 4719, "Pressuremeter Testing in Soils," Vol 04.08. Available from American Society for Testing and Materials, 1916 Race Street, Philadelphia, PA 19103.
31. Canadian Geotechnical Society, 1985. "Canadian Foundation Engineering Manual," 2nd Edition. Available from Canadian Geotechnical Society, BiTech Publishers Ltd., 801 - 1030 W. Georgia Street, Vancouver, B. C. V6E 2Y3.

APPENDIX B

BIBLIOGRAPHY

1. Bachus, R. C. and Barksdale, R. D. 1989. "Design Methodology for Foundations on Stone Columns," *Foundation Engineering: Current Principles and Practices*, Northwestern University, edited by F. H. Kulhawy, Vol 1, pp 244–257, Available from American Society of Civil Engineers, 345 East 47th Street, New York, NY 10017.
2. Baldi, G., Bellotti, R., Ghionna, V., Jamiolkowski, M., Marchetti, S., and Pasqualini, E. 1986. "Flat Dilatometer Tests in Calibration Chambers," *Use of In Situ of In Situ Tests in Geotechnical Engineering*, Geotechnical Special Publication No. 6, pp 431–446, Available from American Society of Civil Engineers, 345 East 47th Street, New York, NY 10017.
3. Barksdale, R. D. and Bachus, R. C. 1983. "Design and Construction of Stone Columns," Vol I, Report No. FHWA/RD-83/026. Available from Federal Highway Administration, Office of Engineering and Highway Operations, Research and Development, Washington D.C. 20590.
4. Binquet, J. and Lee, K. L. 1975. "Bearing Capacity Tests on Reinforced Earth Slabs," *Journal of the Geotechnical Engineering Division*, Vol 101, pp 1241–1276. Available from American Society of Civil Engineers, 345 East 47th Street, New York, NY 10017.
5. Bjerrum, L. 1973. "Problem of Soil Mechanics and Construction on Silt Clays," State-of-the-art Report, *Eighth International Conference on Soil Mechanics and Foundation Engineering*, Vol 3, pp 111–158. Available from USSR National Foundation Engineers, Gosstray USSR, Marx Prospect 12, Moscow K-9.
6. Boussinesq, J. 1885. *Application of Potential to the Study of the Equilibrium and Movements in Elastic Soils*. Available from Gauthier-Villars, 70 rue de Saint-Mandé, F-93100 Montrevil, Paris, France.
7. Bowles, J. E. 1968. *Foundation Analysis and Design*. Available from McGraw Hill Book Co., 1221 Avenue of the Americas, New York, NY 10020.
8. Bowles, J. E. 1988. *Foundation Analysis and Design*, Fourth Edition, Available from McGraw-Hill Book Co., 1221 Avenue of the Americas, New York, NY 10020.
9. Broms, B. B. 1964a. "Lateral Resistance of Piles in Cohesive Soils," *Journal of the Soil Mechanics and Foundations Division*, Vol 90, pp 27–63. Available from American Society of Civil Engineers, 345 East 47th Street, New York, NY 10017.
10. Broms, B. B. 1964b. "Lateral Resistance of Piles in Cohesionless Soil," *Journal of the Soil Mechanics and Foundations Division*, Vol 90, pp 123–156. Available from American Society of Civil Engineers, 345 East 47th Street, New York, NY 10017.
11. Broms, B. B. 1965. "Design of Laterally Loaded Piles," *Journal of the Soil Mechanics and Foundations Division*, Vol 91, pp 79–99. Available from American Society of Civil Engineers, 345 East 47th Street, New York, NY 10017.
12. Brown, J. D. and Meyerhof, G. G. 1969. "Experimental Study of Bearing Capacity in Layered Clays," *Proceedings of the 7th International Conference on Soil Mechanics and Foundation Engineering*, Vol 2, p 45–51. Available from Sociedad Mexicana de Mecanica de Suelos, A. C., Mexico City, Mexico.
13. Bustamante, M. and Gianeselli, L. 1983. "Prezision de la Capacite Portante des Pieux par la Methode Penetrometrique," *Compte Rendu de Recherche S.A.E.R.1.05.022*. Available from Laboratoire Central des Ponts et Chaussees, 58 Boulevard Lefebvre, F-75732, Paris Cedex 15, France.
14. Butler, H. D. and Hoy, H. E. 1977. "Users Manual for the Texas Quick-Load Method for Foundation Load Testing", Report No. FHWA RD-IR,77-8, 59 pp. Available from US Department of Transportation, Federal Highway Administration, Office of Research and Development, Implementation Division, Washington, D. C. 20590.
15. Caquot, A. and Kerisel, J. 1953. "Sur le terme de surface dans le calcul des fondations en milieu pulvérulent", *Third International Conference on Soil Mechanics and Foundation Engineering*, Vol 1, pp 336–337. Available from Organizing Committee ICOSOMEF, Gloriastrasse 39, Zurich 6, Switzerland.
16. Das, B. M. 1987. "Bearing Capacity of Shallow Foundation on Granular Column in Weak Clay",

Foundation Engineering: Current Principles and Practices, Northwestern University, editor F. H. Kulhawy, Vol 2, pp 1252–1263. Available from American Society of Civil Engineers, 345 East 47th Street, New York, NY 10017.

17. Davisson, M. T. 1972. "High Capacity Piles", *Proceedings Lecture Series, Innovations in Foundation Construction,* 52 pp., Illinois Section, American Society of Civil Engineers, New York, NY 10017.
18. DeBeer, E. E. 1965. "The Scale Effect on the Phenomenon of Progressive Rupture in Cohesionless Soils", *Sixth International Conference on Soil Mechanics and Foundation Engineering,* Vol 2, pp 13–17. Available from University of Toronto Press, 63A St. George Street, Toronto, ON M5S1A6, Canada.
19. Edinger, P. H. 1989. "Seismic Response Consideration in Foundation Design", *Foundation Engineering: Current Principles and Practices,* Northwestern University, editor F. H. Kulhawy, Vol 1, pp 814–824, Available from American Society of Civil Engineers, 345 East 47th Street, New York, NY 10017.
20. Edris, E. V., Jr. 1987. "User's Guide: UTEXAS2 Slope Stability Package", Vols I and II, Instruction Report GL-87-1. Available from Research Library, US Army Engineer Waterways Experiment Station, Vicksburg, MS 39180.
21. Gibbs, H. J. and Holtz, W. G. 1957. "Research on Determining the Density of Sands by Spoon Penetration Testing", *Fourth International Conference on Soil Mechanics and Foundation Engineering,* Vol 1, pg 35. Available from Butterworths Publications, Ltd., 88 Kingsway, London, WC2, England.
22. Goble Rausche Likins and Associates, Inc. (GRL) 1988. *GRLWEAP Wave Equation Analysis of Pile Driving.* Available from GRL, 4535 Emery Industrial Parkway, Cleveland, OH 44128.
23. Gurtowski, T. M. and Wu, M. J. 1984. "Compression Load Tests on Concrete Piles in Alluvium", *Analysis and Design of Pile Foundations,* J. R. Meyer, ed., pp. 138–153. Available from American Society of Civil Engineers, 345 East 47th Street, New York, NY 10017.
24. Hanna, A. M. and Meyerhof, G. G. 1980. "Design Charts for Ultimate Bearing Capacity of Foundations on Sand Overlying Soft Clay," *Canadian Geotechnical Journal,* Vol 17, pp 300–303. Available from National Research Council of Canada, Research Journals, Ottawa, ON KlA OR6, Canada.
25. Hansen, J. B. 1963. "Hyperbolic Stress-Strain Response: Cohesive Soils", Discussion, *Journal of the Soil Mechanics and Foundations Division,* Vol 89, No. SM4, pp 241–242, Available from American Society of Civil Engineers, 345 East 47th Street, New York, NY 10017.
26. Hansen, J. B. 1970. "A Revised and Extended Formula for Bearing Capacity", *Danish Geotechnical Institute Bulletin,* No. 28, Available from The Danish Geotechnical Institute, Maglebjergvej 1, DK-2800 Lyngby, Denmark.
27. Herbich, J. B., Schiller, Jr., R. E. and Dunlap, W. A. 1984. "Seafloor Scour: Design Guidelines for Ocrean-Founded Structures," pp 147, 182. Available from Marcel Dekker, Inc., 270 Madison Avenue, New York, NY 10016.
28. Hartman, J. P., Jaeger, J. J., Jobst, J. J., and Martin, D. K. 1989. "User's Guide: Pile Group Analysis (CPGA) Computer Program," Technical Report ITL-89-3. Available from Research Library, US Army Engineer Waterways Experiment Station, Vicksburg, MS 39180.
29. Jaeger, J. J., Jobst, J. J., and Martin, D. K. 1988. "User's Guide: Pile Group Graphics (CPGG) Computer Program," Technical Report ITL-88-2. Available from US Army Engineer Waterways Experiment Station, Vicksburg, MS 39180.
30. Jamiolkowski, M., Ghionna, V. N., Lancellotto, R. and Pasqualini, E. 1988. "New Correlations of Penetration Tests for Design Practice", *Penetration Testing 1988 ISOPT-1,* J. DeRuiter, ed., Vol 1, pp 263–296. Available from A. A. Balkema Publishers, Old Post Road, Brookfield, VT 05036.
31. Johnson, L. D. 1984. "Methodology for Design and Construction of Drilled Shafts in Cohesive Soils", Technical Report GL-84-5, Available from Research Library, US Army Engineer Waterways Experiment Station, Vicksburg, MS 39180.
32. Kraft, L. M., Ray, R. P., and Kagawa, T. 1981. "Theoretical t-z Curves", *Journal of the Geotechnical Engineering Division,* Vol 107, pp 1543–1561. Available from American Society of Civil Engineers, 345 East 47th Street, New York, NY 10017.
33. Leonards, G. A. 1962. *Foundation Engineering,* pp 540–544. Available from McGraw-Hill Bokk Co., 1221 Avenue of the Americas, New York, NY 10020.
34. Lobacz, E. F. 1986. "Arctic and Subarctic Construction: General Provisions", Special Report 86-17. Available from US Army Cold Regions and Engineering Laboratory, Hanover, NH 03755.
35. Mair, R. J. and Wood, D. M. 1987. *Pressuremeter Testing,* pp 51–52. Available from Construction Industry Research and Information Association, London SW1P3AU, England.

36. McKeen, R. G. and Johnson, L. D. 1990. "Climate Controlled Soil Design Parameters for Mat Foundations", *Journal of the Geotechnical Engineering Division*, Vol 116, No. GT7, pp 1073–1094. Available from American Society of Civil Engineers, 345 East 47th Street, New York, NY 10017.
37. Meyerhof, G. G. 1953. "The Bearing Capacity of Foundations Under Eccentric and Inclined Loads', *Third International Conference on Soil Mechanics and Foundation Engineering*, Vol 1, pp 440–445. Available from Organizing Committee ICOSOMEF, Gloriastrasse 39, Zurich 6, Switzerland.
38. Meyerhof, G. G. 1956. "Penetration Tests and Bearing Capacity of Cohesionless Soils," *Journal of the Soil Mechanics and Foundations Division*, Vol 82, No. SM1, pp 866–1 to 866–19. Available from American Society of Civil Engineers, 345 East 47th Street, New York, NY 10017.
39. Meyerhof, G. G. 1963. "Some Recent Research on the Bearing Capacity of Foundations", *Canadian Geotechnical Journal*, Vol 1, No. 1, pp 16–26. Available from National Research Council of Canada, Research Journals, Ottawa, ON K1A OR6, Canada.
40. Meyerhof, G. G. 1974. "Penetration Testing Outside Europe: General Report", *Proceedings of the European Symposium on Penetration Testing*, Vol 2.1, pp 40–48. Available from National Swedish Institute for Building Research, P. O. Box 785, S-801-29-GAVLEÄ, Sweden.
41. Meyerhof, G. G. 1974. "Ultimate Bearing Capacity of Footings on Sand Overlying Clay," *Canadian Geotechnical Journal*, Vol 11, pp 223–229. Available from National Research Council of Canada, Research Journals, Ottawa, ON KlA OR6, Canada.
42. Meyerhof, G. G. 1976. "Bearing Capacity and Settlement of Pile Foundations", *Journal of the Geotechnical Engineering Division*, Vol 102, GT3, pp 197–228, Available from American Society of Civil Engineers, 345 East 47th Street, New York, NY 10017.
43. Meyerhof, G. G. 1983. "Scale Effects of Ultimate Pile Capacity", *Journal of Geotechnical Engineering*, Vol 109, No 6, pp 797–806. Available from American Society of Civil Engineers, 345 East 47th Street, New York, NY 10017.
44. Mosher, R. L. and Pace, M. E. 1982. "User's Guide: Computer Program for Bearing Capacity Analysis of Shallow Foundations (CBEAR)", Instruction Report K-82-7. Available from Research Library, US Army Engineer Waterways Experiment Station, Vicksburg, MS 39180.
45. Osterberg, J. O. 1984. "A New Simplified Method for Load Testing Drilled Shafts", *Foundation Drilling*, August, pp 9–11. Available from Association of Drilled Shaft Contractors, 1700 Eastgate Drive, Suite 208, Garland, TX 75228.
46. Peck, R. B. and Bazarra, A. 1969. "Discussion of Settlement of Spread Footings on Sand," by D'Appolonia, et. a., *Journal of the Soil Mechanics and Foundations Division*, Vol 95, pp 905–909. Available from American Society of Civil Engineers, 345 East 47th Street, New York, NY 10017.
47. Peck, R. B., Hanson, W. E., and Thornburn, T. H. 1974. *Foundation Engineering.* Available from John Wiley and Sons, Ltd., 605 3rd Avenue, New York, NY 10016.
48. Poulos, H. G. 1971a. "Behavior of Laterally Loaded Piles: I-Single Piles", *Journal of Soil Mechanics and Foundation Engineering*, Vol 97, No SM5, pp 711–731. Available from American Society of Civil Engineers, 345 East 47th Street, New York, NY 10017.
49. Poulos, H. G. 1971b. "Behavior of Laterally Loaded Piles: II-Pile Groups", *Journal of Soil Mechanics and Foundation Engineering*, Vol 97, No. SM5, pp 733–757. Available from American Society of Civil Engineers, 345 East 47th Street, New York, NY 10017.
50. Poulos, H. G. and Davis, E. H. 1980. *Pile Foundation Analysis and Design*, p. 26. Available from John Wiley and Sons, Ltd., 605 3rd Avenue, New York, NY 10016.
51. Reese, L. C. 1986. "Behavior of Piles and Pile Groups Under Lateral Load", Publication No. FHWA-RD-85-106. Available from US Department of Transportation, Federal Highway Administration, Office of Implementation, McLean, VA 22101.
52. Reese, L. C., Cooley, L. A. and Radhakrishnan, N. 1984. "Laterally Loaded Piles and Computer Program COM624G", Technical Report K-84-2. Available from Research Library, US Army Engineer Waterways Experiment Station, Vicksburg, MS 39180.
53. Reese, L. C. and O'Neill, M. W. 1988. *Drilled Shafts: Construction Procedures and Design Methods*, Publication No. FHWA-HI-88-042/ADSC-TL-4. Available from US Department of Transportation, Federal Highway Administration, Office of Implementation, McLean, VA 22101 or The Association of Drilled Shaft Contractors, 1700 Eastgate Drive, Suite 208, Garland, TX 75228.
54. Reese, L. C. and Wright, S. J. 1977. "Drilled Shaft Design and Construction Guidelines Manual, Con-

struction of Drilled Shafts and Design for Axial Loading", Vol I. Available from US Department of Transportation, Federal Highway Administration, Office of Implementation, McLean, VA 22101.

55. Robertson, P. K. and Campanella, R. G. 1983. "Interpretation of Cone Penetration Tests - Part I (Sand)", *Canadian Geotechnical Journal*, Vol 20, No 4, pp 718–733. Available from National Research Council of Canada, Research Journals, Ottawa, ON K1A OR6, Canada.
56. Schmertmann, J. H. 1978. *Guidelines for Cone Penetration Test Performance and Design*, Report No. FHWA-TS-78-209. Available from US. Department of Transportation, Federal Highway Administration, Office of Implementation, McLean, VA 22101.
57. Seed, H. B. and Reese, L. D. 1957. "The Action of Soft Clay Along Friction Piles", *Transactions*, Vol 122, pp 731–753. Available from American Society of Civil Engineers, 345 East 47th Street, New York, NY 10017.
58. Skempton, A. W. 1986. "Standard Penetration Test Procedures and The Effects in Sands of Overburden Pressure, Relative Density, Particle Size, Aging and Overconsolidation", *Geotechnique*, Vol 36, No. 3, pp. 425–447. Available from Thomas Telford Ltd., 1-7 Great George Street, Westminster, London, SW1P 3AA, England.
59. Spangler, M. G. and Handy, R. L. 1982. *Soil Engineering*, Fourth Edition. Available from Harper & Row Publishers, 10 East 53rd Street, New York, NY 10022.
60. Snethen, D. R., Johnson, L. D. and Patrick, D. M. 1977. "An Evaluation of Expedient Methodology for Identification of Potentially Expansive Soils," Report No. FHWA-RD-77. Available from Office of Research and Development, Federal Highway Administration, US Department of Transportation, Washington, D.C. 20590.
61. Stewart, J. P. and Kulhawy, F. H. 1981. "Experimental Investigation of the Uplift Capacity of Drilled Shaft Foundations in Cohesionless Soil", Contract Report B-49 (6), Niagara Mohawk Power Corporation, Syracuse, NY 14853. Available as Geotechnical Engineering Report 81-2 from School of Civil and Environmental Engineering, Cornell Unliversity, Ithaca, NY 14850.
62. Strom, R., Abraham, K. and Jones, H. W. 1990. "User's Guide: Pile Group - Concrete Pile Analysis Program (CPGC) Computer Program," Instruction Report ITL-90-2. Available from Research Library, US Army Engineer Waterways Experiment Station, Vicksburg, MS 39180.
63. Tand, K. E., Funegard, E. G., and Briaud, J-L. 1986. "Bearing Capacity of Footngs on Clay CPT Method", *Use of In Situ Tests in Geotechnical Engineering*, Geotechnical Special Publication No. 6, pp 1017–1033. Available from American Society of Civil Engineers, 345 East 47th Street, New York, NY 10017.
64. Terzaghi, K. 1943. "Evaluation of Coefficient of Subgrade Reaction", *Geotecnique*, Vol 5, No. 4, pp 118–143. Available from Thomas Telford Ltd., 1-7 Great George Street, Westminster, London, SW1P 3AA, England.
65. Terzaghi, K. and Peck, R. B. 1967. *Soil Mechanics in Engineering Practice*, Second Edition, pp 117, 265, 341. Available from John Wiley and Sons, Ltd., 605 3rd Avenue, New York, NY 10016.
66. Tokimatsu, K. and Seed, H. B. 1984. "Simplified Procedures for the Evaluation of Settlements in Sands Due to Earthquake Shaking", Report No. UCB/EERC-84/16. Available from Earthquake Engineering Research Center, University of California, Berkeley, CA 94720.
67. Tokimatsu, K. and Seed, H. B. 1987. "Evaluation of Settlements in Sands Due to Earthquake Shaking", *Journal of Geotechnical Engineering*, Vol 113, pp 861–878. Available from American Society of Civil Engineers, 345 East 47th Street, New York, NY 10017.
68. Tomlinson, M. J. 1980. *Foundation Design and Construction*, Fourth Edition, Available from Pitman Publishing Limited, 128 Long Acre, London WC2E 9AN, UK.
69. Vanikar, S. N. 1986. "Manual on Design and Construction of Driven Pile Foundations", Report No. FHWA-DP-66-1. Available from US Department of Transportation, Federal Highway Administration, Demonstration Projects Division, HHO-33, 400 7th Street SW, Washington, D.C. 20590.
70. Vesic, A. S. 1963. "Bearing Capacity of Deep Foundations in Sand", Highway Research Record 39, pp 112–153. Available from National Academy of Sciences, 2101 Constitution Avenue NW, Washington, D.C. 20418.
71. Vesic, A. S. 1969. "Discussion: Effects of Scale and Compressibility on Bearing Capacity of Surface Foundations", *Seventh International Conference on Soil Mechanics and Foundation Engineering*, Vol 3, pp 111–159. Available from Sociedad Mexicana de Mecanica de Suelos, A. C., Mexico City, Mexico.
72. Vesic, A. S. 1971. "Breakout Resistance of Object Embedded in Ocean Bottom", *Journal of the Soil Medchanics and Foundation Division*, Vol 97, No. SM9, pp 1183–1205. Available from American

Society of Civil Engineers, 345 East 47th Street, New York, NY 10017.

73. Vesic, A. S. 1973. "Analysis of Ultimate Loads of Shallow Foundations", *Journal of Soil Mechanics and Foundations Division*, Vol 99, No. SM1, pp 45–73. Available from American Society of Civil Engineers, 345 East 47th Street, New York, NY 10017.
74. Vesic, A. S. 1975. "Bearing Capacity of Shallow Foundations", *Foundation Engineering Handbook*, pp 121–147, editors H. F. Winterkorn and H. Y. Fang. Available from Van Nostrand Reinhold Company, 115 5th Avenue, New York, NY 10003.
75. Vesic, A. S. 1977. "Design of Pile Foundations", *National Cooperative Highway Research Program Synthesis of Highway Practice*, No. 42. Available from Transportation Research Board, 2101 Constitution Avenue, Washington, D. C. 20418.
76. Vijayvergiya, V. N. and Focht, J. A., Jr. 1972. "A New Way to Predict Capacity of Piles in Clay", No. OTC Paper 1718, *Fourth Offshore Technology Conference*, Houston, TX 75206. Available from Offshore Technology Conference, P.O. Box 833868, Richardson, TX 75080.
77. Vijayvergiya, V. N. 1977. "Load-Movement Characteristics of Piles, *Port 77 Conference*, Long Beach, CA. Available from American Society of Civil Engineers, 345 East 47th Street, New York, NY 10017.
78. Wang, M. C.,Yoo, C. S. and Hsieh, C.W. 1987. "Effect of Void on Footing Behavior Under Eccentric and Inclined Loads", *Foundation Engineering: Current Principles and Practices*, Vol 2, pp 1226–1239, Northwestern University, editor F. H. Kulhawy, American Society of Civil Engineers, 345 East 47th Street, New York, NY 10017.

APPENDIX C

COMPUTER PROGRAM AXILTR

C-1. Organization

Program AXILTR, AXIal Load-TRansfeR, consists of a main routine and two subroutines. The main routine feeds in the input data, calculates the effective overburden stress, and determines whether the load is axial down-directed, pullout, or if uplift/downdrag forces develop from swelling or consolidating soil. The main routine also prints out the computations. Subroutine BASEL calculates the displacement at the base for given applied down-directed loads at the base. Subroutine SHAFL evaluates the load transferred to and from the shaft for relative displacements between the shaft and soil. An iteration scheme is used to cause the calculated applied loads at the top (butt) to converge within 10 percent of the input load applied at the top of the shaft.

A. INPUT DATA. Input data are illustrated in Table C-1 with descriptions given in Table C-2.

1. The program is set to consider up to a total of 40 soil types and 100 soil elements. Figure C-1 provides an example layout of soil types and elements used in AXILTR.

2. The program can accommodate up to 18 points of the load-displacement curve. This capacity may be altered by adjusting the PARAMETER statement in the program.

3. The input data are placed in a file, "DATLTR.TXT". These data are printed in output file, "LTROUT.TXT" illustrated in Table C-3a.

B. OUTPUT DATA. Results of the computations by AXILTR are printed in LTROUT.TXT illustrated in Table C-3b. Table C-3c provides a description of calculations illustrated in Table C-3b.

1. Load-displacement data are placed in file LDCOM.DAT for plotting by graphic software.

2. Load-depth data for a given applied load on the pile top are placed in file LDEP.DAT for plotting by graphic software.

3. Displacement-depth data for a given applied load on the pile top are placed in file MDEP.DAT for plotting by graphic software.

C-2. Applications

The pullout, uplift and downdrag capabilities of AXILTR are illustrated by two example problems. The accuracy of these solutions can be increased by using more soil layers, which increases control over soil input parameters such as swell pressure, maximum past pressure, and shear strength.

A. PULLOUT AND UPLIFT. Table C-4 illustrates input data required to determine performance of a 2 ft diameter drilled shaft 50 ft long constructed in an expansive clay soil of two layers, NMAT = 2. The shaft is underreamed with a 5-ft diameter bell. Soil beneath the shaft is nonexpansive. The shaft is subject to a pullout force of 300 kips. Refer to Figure C-1 for a schematic representation of this problem.

1. Bearing Capacity. The alpha skin friction and local shear base capacity models are selected. Option to input the reduction factor α (I = 6) was used. The selected α's for the two soils is 0.9. A high was selected because expansive soil increases pressure against the shaft, which may raise the skin friction.

2. Load-Transfer Models. The Kraft, Ray, and Kagawa skin friction and the Vijayvergiya base load-transfer models (K = 8) were selected. Two points for the elastic modulus of the shaft concrete were input into the program.

3. Results. The results are plotted in Figure C-2 for a pullout force of 300,000 pounds. Results of the computation placed in file "LTROUT.TXT" are shown in Table C-5.

a. Total and base ultimate bearing capacity is about 1,200 and 550 kips, respectively, Figure C-2a. Base and total capacity is 250 and 600 kips, respectively, if settlement is limited to 0.5 inch, which is representative of a FS of approximately 2.

b. The distribution of load with depth, Figure C-2b, is a combination of the shapes indicated in Figures 5-9 and 5-10 because both pullout and uplift forces must be resisted.

Table C-1. Input Data

Line	Input Parameters	Format Statement
1	TITLE	20A4
2	NMAT NEL DX GWL LO IQ IJ	2I5,2F6.2,3I5
3	I J K SOILP DS DB	3I5,3F10.3
4	E50 (Omitted unless K = 2, 5, 9)	E13.3
5	LLL	I5
6	MAT GS EO WO PS CS CC C PHI AK PM (Line 5 repeated for each material M = 1,NMAT)	I3,3F6.2,F7.0, 2F7.2,F7.0,2F6.2 F7.0
7	ALPHA (Omitted unless I = 6) (α input for each material MAT = 1,NMAT)	7F10.5
8	M IE(M) (Line 8 repeated for each element M and number of soil IE(M). Start with 1. The last line is NEL NMAT)	2I5
9	RFF GG (Omitted unless K = 7, 8, 9)	F6.3,E13.3
10	(Omitted unless K = 3,4,5,6)	
10a	NCA (<12)	I5
10b	T(M,1)... T(M,11) (Input for each curve M=1,NCA)	11F6.2
10c	S(M) (Input on new line for each M = 2,11; S(1) input in program as 0.00)	F6.3
11	(Omitted unless I = 5)	
11a	NCC (<12)	I5
11b	FS(N) ZEPP(N) NCUR (Input on new line for each N = 1,NCC)	2F10.3,I5
12	(Omitted unless J = 0)	
12a	NC (>1)	I5
12b	EP(M) ZEP(M) (Input on new line for each M = 1,NC; at least a top and bottom term required)	E13.3,F6.2
13	R(M) S(M) (Omitted unless K = 6; repeat on new line for each M = 1,IJ)	F10.5,F15.3
14	STRUL SOILP XA	3F15.2
15	NON (Omitted unless XA < 0.0)	I5

c. The shaft will heave approximately 0.7 inch, while the soil heaves more than 11 inches at the ground surface, Figure C-2c.

B. DOWNDRAG. Table C-6 illustrates input data required to solve for the performance of the same drilled shaft and soil described in the previous example problem, but the soil is wetter with a much lower swell pressure. Soil shear strength is assumed not to change significantly from the previous example. This shaft is subject to a 150-kip load in addition to the downdrag forces from the settling soil.

1. Bearing Capacity. The alpha skin friction and local shear bearing capacity models are selected similar to the previous example. Option to input the reduction factor α (I = 6) was used. The selected α's are 0.55 and 0.3 for the surface and deeper soils, respectively.

2. Load-Transfer Models. The Seed and Reese skin friction and Reese and Wright base load-transfer models were selected (K = 2). Two points for the elastic modulus of the shaft concrete were input into the program.

3. Results. The results are plotted in Figure C-3 for a downward applied load of 150 kips. Results of the computation placed in file LTROUT.TXT are illustrated in Table C-7.

a. Total and base ultimate bearing capacity,

Table C-2. Description of Input Parameters

Line	Parameter	Description	
1	TITLE	Name of problem	
2	NMAT	Total number of materials	
	NEL	Total number of elements	
	DX	Thickness of each element, ft (usually 0.5 or 1.0 ft)	
	GWL	Depth to groundwater level, ft	
	LO	Amount of output data = 0 Extensive data output used to check the program = 1 Shaft load-displacement behavior and detailed load distribution-displacement response along shaft length for input top load prior to and following soil movement (load transfer, load, shaft compression increment, and shaft movement at given depth = 2 Shaft load-displacement behavior and load distribution-displacement response along shaft length for input top load prior to and following soil movement = 3 Shaft load-displacement behavior and load distribution-displacement response along shaft length for input top load on shaft following soil movement	
	IQ	Total number of shaft increments (shaft length/element thickness); top of shaft at ground surface	
	IJ	Number of points for shaft load-displacement behavior (usually 12, but maximum of 18 for PARAMETER statement = 40)	
3	I	Magnitude of reduction factor α applied to total (undrained) or effective (drained) shear strength for skin friction resistance = 0 $\alpha = 1$ (usually used for drained strength) = 1 $\alpha = \sin(\pi x/L)$, x = depth, ft; L = shaft length, ft = 2 $\alpha = 0.6$ = 3 $\alpha = 0.45$ = 4 $\alpha = 0.3$ = 5 Permits maximum skin friction f_{ss} input as a function of depth, psf (see line 11) = 6 α is input for each material (see line 7)	
	J	Option for elastic shaft modulus = 0 shaft modulus input = 1 shaft modulus set to near infinity	
	K	Option for load-transfer functions (see Figure 5-12)	
		Base	Shaft
		= 0 Consolidation	Seed and Reese
		= 1 Vijayvergiya	Seed and Reese
		= 2 Reese and Wright	Seed and Reese
		= 3 Consolidation	Input (see line 10)
		= 4 Vijayvergiya	Input (see line 10)
		= 5 Reese and Wright	Input (see line 10)
		= 6 Input (see line 13)	Input (see line 10)
		= 7 Consolidation	Kraft, Ray, and Kagawa
		= 8 Vijayvergiya	Kraft, Ray, and Kagawa
		= 9 Reese and Wright	Kraft, Ray, and Kagawa
	SOILP	Pressure on top layer of soil exerted by surrounding structure, fill, etc., psf	
	DS	Diameter shaft, ft	
	DB	Diameter base, ft	

Table C-2. Continued

Line	Parameter	Description
4	E50	Strain at 1/2 maximum deviator stress, Equation 5-19
5	LLL	Option for type of shear failure at base = 0 Local shear failure, Equation 5-7 or $N_c = 7$ = 1 General shear failure, Equation 5-8 or $N_c = 9$
6	MAT	Number of material
	GS	Specific gravity
	EO	Initial void ratio
	WO	Initial water content, percent
	PS	Swell pressure, psf
	CS	Swell index
	CC	Compression index
	C	Cohesion, psf; = undrained strength for total stress analysis; effective cohesion c′ or zero for effective stress analysis
	PHI	Angle of shearing resistance ϕ; = 0 for total stress analysis
	AK	Coefficient of lateral earth pressure
	PM	Maximum past pressure, psf (program sets PM = PS if PM input < PS)
7	ALPHA	Reduction factor α_a for each material MAT, Equation 5-11, Table 5-1, Table 5-10; used when option I = 6, Line 3
8	M	Number of element
	IE(M)	Material number of soil, MAT
9	RFF	Hyperbolic reduction factor R for Kraft, Ray, and Kagawa model, Equation 5-19a; use 1.0 if not known
	GG	Shear modulus G_s, psf, Equation 5-19b
10		Input data for shaft load-transfer curves (K = 3,4,5,6)
10a	NCA	Total number of shaft load-transfer curves to input, < 12
10b	T(M,1) . . . . . . T(M,11)	Skin friction ratio of developed shear strength/maximum mobilized shear strength of each shaft load-transfer curve; 11 values required for each load-transfer curve, the first value T(1,1) = 0.0
10c	S(M)	Movement in inches for all of the T(M,1) . . . T(M,11) curves; only 10 values required from S(2) . . . S(11); S(1) = 0.0 in code; if S(M) in the code is okay (0.0,0.05,0.1,0.15,0.2,0.23,0.3,0.45,0.75,1.05,1.5 inches)
11		Input data for maximum skin friction as a function of depth
	NCC	Total number of maximum skin friction terms to input, <12; program interpolates maximum skin friction between depths
11a	FS(N)	Maximum skin friction $f_{\bar{s}}$ for point N, psf
11b	ZEPP(N)	Depth for the maximum skin friction for point N, ft
11c	NCUR	Number of the shaft load-transfer curve input M in line 10; applicable to the maximum skin friction for point N (Repeat 11a,11b,11c for each N = 1,NCC)
12		Input data for shaft elastic modulus as function of depth; program interpolates the elastic modulus between depths
	NC	Total number of terms of elastic modulus and depth, > 1
12a	EP(M)	Elastic modulus of shaft at point M, psf
12b	ZEP(M)	Depth for the elastic modulus of shaft at point M, ft (An elastic modulus and depth term are required at least at the top and bottom of the shaft)
13		Input data for base displacements if K = 6 (The number of input terms or R(M) and S(M) equals IJ − 1, line 2)
13a	R(M)	Base displacement, in. (The first displacement is 0.0 inches and already input in the program)

Table C-2. Concluded

Line	Parameter	Description
13b	S(M)	Base load for displacement R(M), pounds; the base load for 0.0 displacement is approximated as the overlying soil weight and already input in the program.
14		Structural load, pressure on adjacent soil at the ground surface, and depth of the active zone for heave input for each problem for evaluation of specific load distribution-displacement computations
14a	STRUC	Structural vertical load on top of shaft, pounds
14b	SOILP	Pressure on top layer of soil exerted by surrounding structure, fill, etc., psf
14c	XA	Depth of the active zone for heave, ft; = 0.01 yields load-displacement behavior for zero soil movement; a saturated soil profile is assumed when computing soil movement; < 0.0 program goes to line 15 below
15	NON	Execution stops if 0; program goes to line 1 if > 0

Figure C-3a, is about 550 and 880 kips, respectively. Base and total capacity is about 200 and 500 kips, respectively, if settlement is limited to 0.5 inch. The factor of safety is approximately 1.8 relative to total bearing capacity. The program does not add the vertical plunging failure lines to the curves in Figure C-3a, which leaves the calculated displacement load relationships nearly linear.

b. The distribution of load with depth, Figure C-3b, is representative of downdrag indicated in Figure 5-11. The load on the shaft base is nearly 300 kips or double the applied load at the ground surface.

c. The shaft will settle approximately 1 inch, while the soil settles about 2 inches at the ground surface, Figure C-3c. The soil is heaving near the ground surface, which counters the settlement from downdrag. Maximum settlement is about 3.5 inches at 10 ft of depth.

C-3. Listing

A Fortran listing of the computer program is provided in Table C-8 in case modifications may be required for specific applications.

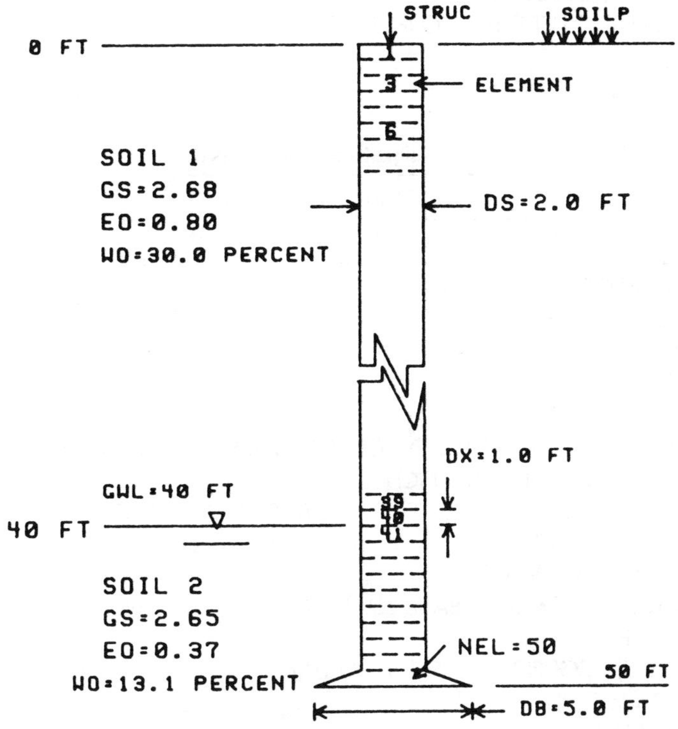

Figure C-1. Schematic diagram of soil/pile elements

Table C-3. Output Data

a. Repeat of Input Data (See Table C-1)

Line	Output Parameters	Fortran Statement
1	TITLE	20A4
2	NMAT= NEL= DX= FT GWL= FT	I5,I5,F6.2,F6.2
	LO= IQ (SHAFT INC)= IJ (NO.LOADS)=	I5,I5,I5
3	I= J= K= SOILP= PSF	I5,I5,I5,F10.2
	DS= FT	F10.2
	DB= FT	F10.2
4	(If K = 2,5,9)	
	E50	E13.3
5	LOCAL SHEAR FAILURE AT BASE - LLL = 0 or	I5
	GENERAL SHEAR FAILURE AT BASE - LLL = 1	I5
6	MAT GS EO WO(%) PS(PSF) CS CC CO(PSF) PHI K PM(PSF)	I3,3F6.2,F7.0, 27.2,F7.0,2F6.2, F7.0
7	(If I = 6) ALPHA =	2(7F10.5)
8	ELEMENT NO OF SOIL	I5,10X,I5
9	(If K = 7,8,9)	
	REDUCTION FACTOR= SHEAR MODULUS=	F6.3,3X,E13.3
10	(If K = 3,4,5,6)	
	NO. OF LOAD-TRANSFER CURVES(<12)?=	I5
	For each curve 1 to NCA:	
	CURVE	I5
	RATIO SHR DEV,M=1,11 ARE	11F6.3
	MOVEMENT(IN.) FOR LOAD TRANSFER M= IS INCHES	I5,F6.3
11	(If I = 5)	
	NO OF SKIN FRICTION-DEPTH TERMS(<12)? ARE	I5
	SKIN FRICTION(PSF) DEPTH(FT) CURVE NO	F10.3,F10.3,I5
12	(If J = 0)	
	E SHAFT(PSF) AND DEPTH(FT):	4(E13.3,2X,F6.2)
13	(If K = 6)	
	BASE DISPLACEMENT(IN.),BASE LOAD(LBS) > FOR POINTS	F10.2,I5

b. *Output Calculations*

Item	Program Prints	Format Statement
1	BEARING CAPACITY= POUNDS	F13.2
2	DOWNWARD DISPLACEMENT	
3	(Omitted unless LO = 0,1)	
	POINT BEARING(LBS)=	F13.2
4	(Omitted unless LO = 0,1)	
	DEPTH LOAD TRANS TOTAL LOAD COM OF INCR TOTAL MOVMT	
	ITER FT LBS LBS INCHES INCHES	5E13.5,I5
5	TOP LOAD TOP MOVEMENT BASE LOAD BASE MOVEMENT	
	LBS INCHES LBS INCHES	4E13.5
6	NEGATIVE UPWARD DISPLACEMENT	
7	TOP LOAD TOP MOVEMENT BASE LOAD BASE MOVEMENT	
	LBS INCHES LBS INCHES	4E13.5
8	STRUC LOAD(LBS) SOILP(PSF) ACTIVE DEPTH(FT)	
	(Line 14 of Table C-2)	F10.0,2F10.2
9	BELL RESTRAINT(LBS)=	F13.2

Table C-3. Continued

Line	Output Parameters	Fortran Statement
10	(If STRUL < 0.0 See Line 14, Table C-2) FIRST ESTIMATE OF PULLOUT RESTRAINT(LBS)=	F13.2
11	LOAD-DISPLACEMENT BEHAVIOR	
12	(If LO <2) EFFECTS OF ADJACENT SOIL	
13	INITIAL BASE FORCE(LBS)= (If LO = 0) BASE FORCE(LBS)=	F13.2
14	DISPLACEMENT(INCHES)= FORCE= POUNDS	F8.4,F12.2
15	ITERATIONS=	I5
16	DEPTH(FT) LOAD(LBS) SHAFT MVMT(IN) SOIL MVMT(IN)	F7.2,2X,E13.5, 2F15.5

c. Description of Calculations

Item	Program Prints	Description
1	BEARING CAP . . .	End bearing capacity, pounds
2	DOWNWARD DISPL	Load-displacment Behavior for zero soil movement in downward direction for IJ points
3	POINT BEARING	Load at bottom of shaft prior to shaft load-transfer calculation, pounds
4	DEPTH	Depth, ft
	LOAD TRANS	Load transferred at given depth along shaft, pounds
	TOTAL LOAD	Total load on shaft at given depth, pounds
	COM OF INCR	Incremental shaft compression at given depth, in.
	TOTAL MOVMT	Shaft-soil relative movement at given depth, in.
	ITER	Number of iterations to complete calculation
5	TOP LOAD	Load at top of shaft, pounds
	TOP MOVEMENT	Displacement at top of shaft, in.
	BASE LOAD	Load at bottom of shaft, pounds
	BASE MOVEMENT	Displacement at bottom of shaft, in.
6	NEGATIVE UPWARD	Load-displacement Behavior for zero soil movement in upward direction for IJ points
7	Same as item 5	
8	STRUC LOAD(LBS)	Load applied on top of shaft, pounds
	SOILP(PSF)	Pressure applied on top of adjacent soil, psf
	ACTIVE DEPTH	Depth of soil beneath ground surface subject to soil heave, ft
9	BELL RESTRAINT	Restraining resistance of bell, pounds
10	FIRST ESTIMATE	Initial calculation of pullout resistance prior to iterations for structural loads less than zero, pounds
11	LOAD-DISPLACE	Load-shaft movement distribution for given structural load
12	EFFECTS OF ADJ	Effects of soil movement considered in load-displacement behavior
13	INITIAL BASE	Initial calculation of force at bottom of shaft prior to iterations
14	DISPLACEMENT	Displacement at bottom of shaft after 100 iterations, in.
	FORCE=	Force at bottom of shaft, pounds after 100 iterations, pounds

Table C-3. Concluded

Item	Program Prints	Description
15	ITERATIONS	Total number of iterations to converge to solution
16	DEPTH(FT)	Depth, ft
	LOAD(LBS)	Load at given depth, pounds
	SHAFT MVMT(IN)	Shaft displacement, in.
	SOIL MVMT(IN)	Soil movement, in.

Table C-4. Listing of Data Input for Expansive Soil, File DATLTR.TXT

```
   EXPANSIVE SOIL
 2  50  1.0  40.  2 50  16
 6  0  8  0.0  2.0  5.00
 0
1  2.68  .8   30.   4800.  .1  .2  2000.  .0   .7   7000.
2  2.65  .37  13.1  6000.  .1  .2  4000.  .0  2.   10000.
 0.9          0.9
  1  1
41  2
50  2
.900  1.600E+05
 2
 4.333E 08   .0
 4.333E 08 50.0
-300000.   .0   50.
       0.   .0  -1.0
 0
```

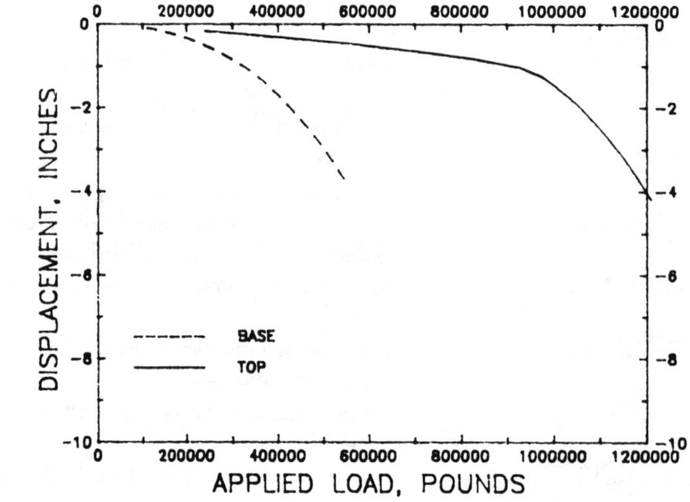

a. LOAD-DISPLACEMENT BEHAVIOR, FILE LDCOM

Figure C-2. Plotted output for pullout and uplift problem

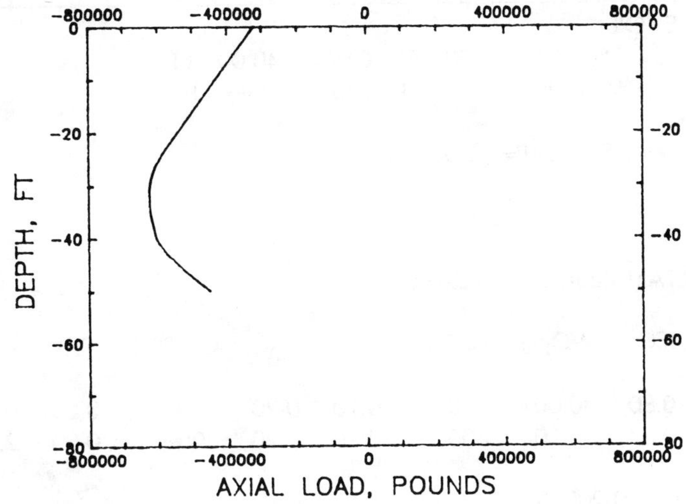

b. AXIAL LOAD–DEPTH BEHAVIOR, FILE LDEP

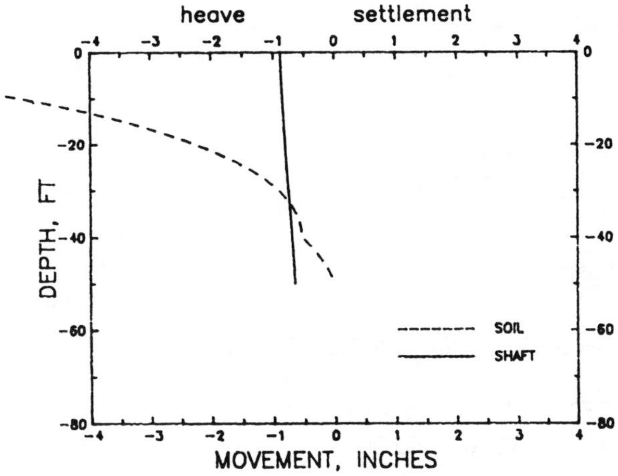

c. DISPLACEMENT–DEPTH BEHAVIOR, FILE MDEP

Figure C-2. Concluded

Table C-5. Listing of Output, File LTROUT.TXT

```
     EXPANSIVE SOIL
NMAT=  2  NEL=  50  DX=  1.00  FT  GWL=  40.00  FT
  LO=  2  IQ  (SHAFT  INC)=  50  IJ  (NO.LOADS)=  16

I=  6  J=  0  K=  8  SOILP=  0.00  PSF
DS= 2.00 FT
DB= 5.00 FT

LOCAL SHEAR FAILURE AT BASE - LLL= 0

MAT   GS    EO    WO(%)   PS(PSF)   CS    CC    CO(PSF)   PHI    K     PM(PSF)

1     2.68  0.80  30.00   4800.     0.10  0.20  2000.     0.00   0.70   7000.
2     2.65  0.37  13.10   6000.     0.10  0.20  .4000.    0.00   2.00  10000.

ALPHA= 0.90000  0.90000

      ELEMENT           NO OF SOIL
         1                  1
         2                  1
         .                  1
         .                  1
        40                  1
        41                  2
        42                  2
         .                  2
         .                  2
        50                  2

REDUCTION FACTOR= 0.900   SHEAR MODULUS= 0.160E+06

E SHAFT(PSF) AND DEPTH(FT):
  0.433E+09  0.00  0.433E+09  50.00

  BEARING CAPACITY= 549778.69 POUNDS

DOWNWARD DISPLACEMENT
```

TOP LOAD LBS	TOP MOVEMENT INCHES	BASE LOAD LBS	BASE MOVEMENT INCHES
0.24017E+06	0.17714E+00	0.10946E+06	0.99065E-01
0.34507E+06	0.26781E+00	0.13882E+06	0.15855E+00
0.45773E+06	0.37719E+00	0.16817E+06	0.23526E+00
0.58421E+06	0.50996E+00	0.19753E+06	0.33139E+00
0.71040E+06	0.66509E+00	0.22688E+06	0.44915E+00
0.82982E+06	0.84256E+00	0.25624E+06	0.59070E+00
0.92817E+06	0.10432E+01	0.28559E+06	0.75826E+00
0.97601E+06	0.12587E+01	0.31494E+06	0.95401E+00
0.10054E+07	0.14978E+01	0.34430E+06	0.11801E+01
0.10347E+07	0.17694E+01	0.37365E+06	0.14388E+01
0.10641E+07	0.20758E+01	0.40301E+06	0.17323E+01
0.10934E+07	0.24192E+01	0.43236E+06	0.20627E+01

Table C-5. Continued

TOP LOAD LBS	TOP MOVEMENT INCHES	BASE LOAD LBS	BASE MOVEMENT INCHES
0.11228E+07	0.28017E+01	0.46172E+06	0.24323E+01
0.11521E+07	0.32256E+01	0.49107E+06	0.28432E+01
0.11815E+07	0.36930E+01	0.52042E+06	0.32977E+01
0.12108E+07	0.42061E+01	0.54978E+06	0.37979E+01

NEGATIVE UPWARD DISPLACEMENT

TOP LOAD LBS	TOP MOVEMENT INCHES	BASE LOAD	BASE MOVEMENT INCHES
−0.18590E+05	−0.37138E-02	0.00000E+00	0.00000E+00
−0.31134E+05	−0.16708E-01	0.00000E+00	−0.10000E-01
−0.43689E+05	−0.29706E-01	0.00000E+00	−0.20000E-01
−0.68793E+05	−0.55704E-01	0.00000E+00	−0.40000E-01
−0.11899E+06	−0.10770E+00	0.00000E+00	−0.80000E-01
−0.21806E+06	−0.21160E+00	0.00000E+00	−0.16000E+00
−0.38024E+06	−0.41089E+00	0.00000E+00	−0.32000E+00
−0.61240E+06	−0.78911E+00	0.00000E+00	−0.64000E+00
−0.69610E+06	−0.14531E+01	0.00000E+00	−0.12800E+01
−0.69610E+06	−0.27331E+01	0.00000E+00	−0.25600E+01
0.69610E+06	−0.52931E+01	0.00000E+00	−0.51200E+01
−0.69610E+06	−0.10413E+02	0.00000E+00	−0.10240E+02
−0.69610E+06	−0.20653E+02	0.00000E+00	−0.20480E+02
−0.69610E+06	−0.41133E+02	0.00000E+00	−0.40960E+02
−0.69610E+06	−0.82093E+02	0.00000E+00	−0.81920E+02
−0.69610E+06	−0.16401E+03	0.00000E+00	−0.16384E+03

STRUC LOAD(LBS)	SOILP(PSF)	ACTIVE DEPTH(FT)
−300000.	0.00	50.00

BELL RESTRAINT(LBS)= 449915.44

FIRST ESTIMATE OF PULLOUT RESTRAINT(LBS)= 541894.31

LOAD-DISPLACEMENT BEHAVIOR

INITIAL BASE FORCE(LBS)= −788275.25

DISPLACEMENT(INCHES)= −0.2475 FORCE= −667768.19 POUNDS

DISPLACEMENT(INCHES)= −0.4975 FORCE= −532357.44 POUNDS

DISPLACEMENT(INCHES)= −0.6525 FORCE= −449443.94 POUNDS

ITERATIONS= 262

DEPTH(FT)	LOAD(LBS)	SHAFT MVMT(IN)	SOIL MVMT(IN)
0.00	−0.32427E+06	−0.88276	−11.94514
1.00	−0.33520E+06	−0.87985	−10.67843
2.00	−0.34613E+06	−0.87685	−9.72980

Table C-5. Concluded

DEPTH(FT)	LOAD(LBS)	SHAFT MVMT(IN)	SOIL MVMT(IN)
3.00	−0.35706E+06	−0.87375	−8.92906
4.00	−0.36799E+06	−0.87055	−8.22575
5.00	−0.37892E+06	−0.86726	−7.59519
6.00	−0.38985E+06	−0.86387	−7.02274
7.00	−0.40078E+06	−0.86039	−6.49865
8.00	−0.41171E+06	−0.85681	−6.01600
9.00	−0.42264E+06	−0.85313	−5.56958
10.00	−0.43357E+06	−0.84936	−5.15537
11.00	−0.44450E+06	−0.84549	−4.77014
12.00	−0.45543E+06	−0.84152	−4.41124
13.00	−0.46636E+06	−0.83746	−4.07648
14.00	−0.47729E+06	−0.83330	−3.76401
15.00	−0.48822E+06	−0.82904	−3.47223
16.00	−0.49915E+06	−0.82469	−3.19976
17.00	−0.51008E+06	−0.82024	−2.94538
18.00	−0.52101E+06	−0.81570	−2.70805
19.00	−0.53194E+06	−0.81105	−2.48680
20.00	−0.54287E+06	−0.80632	−2.28080
21.00	−0.55380E+06	−0.80148	−2.08927
22.00	−0.56473E+06	−0.79655	−1.91153
23.00	−0.57566E+06	−0.79153	−1.74696
24.00	−0.58613E+06	−0.78641	−1.59498
25.00	−0.59556E+06	−0.78120	−1.45506
26.00	−0.60381E+06	−0.77591	−1.32673
27.00	−0.61073E+06	−0.77056	−1.20953
28.00	−0.61621E+06	−0.76515	−1.10306
29.00	−0.62027E+06	−0.75970	−1.00692
30.00	−0.62304E+06	−0.75422	−0.92078
31.00	−0.62444E+06	−0.74872	−0.84428
32.00	−0.62465E+06	−0.74321	−0.77713
33.00	−0.62386E+06	−0.73771	−0.71902
34.00	−0.62223E+06	−0.73222	−0.66969
35.00	−0.61992E+06	−0.72674	−0.62887
36.00	−0.61710E+06	−0.72129	−0.59633
37.00	−0.61390E+06	−0.71587	−0.57183
38.00	−0.61049E+06	−0.71047	−0.55516
39.00	−0.60701E+06	−0.70510	−0.54610
40.00	−0.60360E+06	−0.69977	−0.54447
41.00	−0.59487E+06	−0.69448	−0.46514
42.00	−0.58401E+06	−0.68929	−0.39155
43.00	−0.57119E+06	−0.68420	−0.32363
44.00	−0.55675E+06	−0.67922	−0.26128
45.00	−0.54103E+06	−0.67439	−0.20443
46.00	−0.52416E+06	−0.66969	−0.15300
47.00	−0.50642E+06	−0.66515	−0.10692
48.00	−0.48799E+06	−0.66077	−0.06611
49.00	−0.46897E+06	−0.65655	−0.03049
50.00	−0.44944E+06	−0.65250	0.00000

STRUC LOAD(LBS)	SOILP(PSF)	ACTIVE DEPTH(FT)
0.	0.00	−1.00

Table C-6. Listing of Data Input for Settling Soil, File DATLTR.TXT

```
    SETTLING SOIL
  2  50  1.0  40.  2  50  16
  6  0  2  0.0  2.0  5.00
   0.010
  0
1  2.68  .8  30.  1200.  .05  .1 2000.  .0  .7  4000.
2  2.65  .37  13.1  6000.  .05  .1  4000.  .0 2.  10000.
 0.55  0.3
  1  1
 41  2
 50  2
  2
  4.333E 08      .0
  4.333E 08   50.0
150000.  .0 50.
     0.  .0  -1.0
  0
```

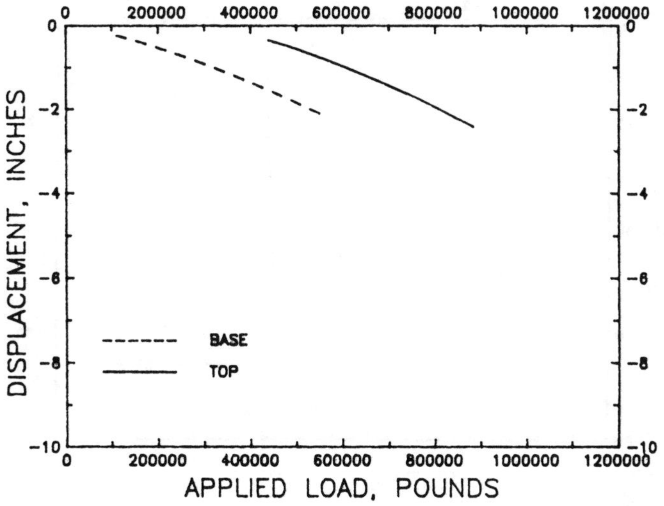

a. LOAD-DISPLACEMENT BEHAVIOR, FILE LDCOM

Figure C-3. Plotted output for downdrag problem

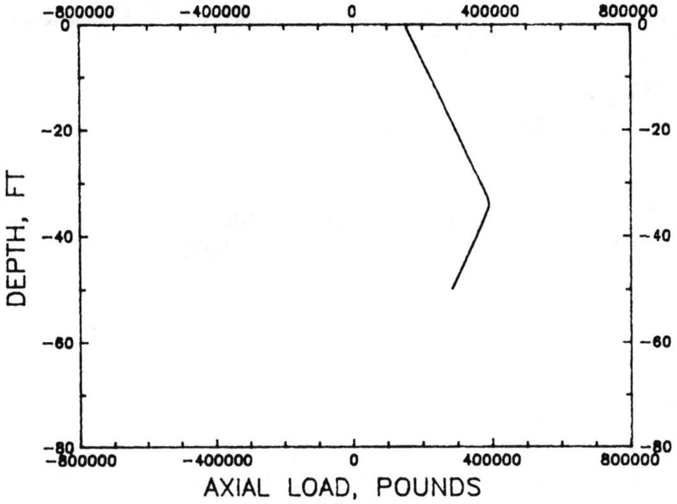

b. AXIAL LOAD-DEPTH BEHAVIOR, FILE LDEP

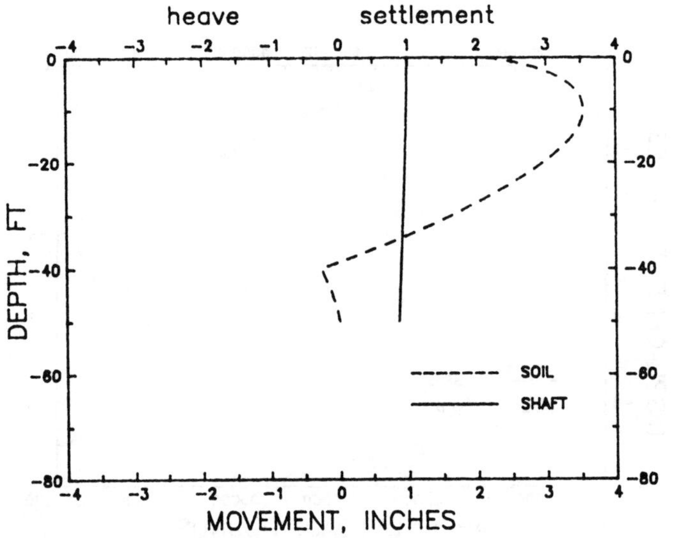

c. DISPLACEMENT-DEPTH BEHAVIOR, FILE MDEP

Figure C-3. Concluded

Table C-7. Listing of Output, File LTROUT.TXT

```
    SETTLING SOIL
NMAT= 2  NEL= 50  DX= 1.00 FT  GWL= 40.00 FT
  LO= 2  IQ (SHAFT INC)= 50  IJ (NO.LOADS)= 16

I= 6  J= 0  K= 2  SOILP= 0.00 PSF
DS= 2.00 FT
DB= 5.00 FT

E50= 0.100E-01

LOCAL SHEAR FAILURE AT BASE - LLL= 0
```

MAT	GS	EO	WO(%)	PS(PSF)	CS	CC	CO(PSF)	PHI	K	PM(PSF)
1	2.68	0.80	30.00	1200.	0.05	0.10	2000.	0.00	0.70	4000.
2	2.65	0.37	13.10	6000.	0.05	0.10	4000.	0.00	2.00	10000.

ALPHA= 0.55000 0.30000

ELEMENT	NO OF SOIL
1	1
2	1
.	1
.	1
40	1
41	2
42	2
.	2
.	2
50	2

E SHAFT(PSF) AND DEPTH(FT):
0.433E+09 0.00 0.433E+09 50.00

BEARING CAPACITY= 549778.69 POUNDS

DOWNWARD DISPLACEMENT

TOP LOAD LBS	TOP MOVEMENT INCHES	BASE LOAD LBS	BASE MOVEMENT INCHES
0.43825E+06	0.36209E+00	0.10946E+06	0.24071E+00
0.47316E+06	0.46787E+00	0.13882E+06	0.33163E+00
0.50252E+06	0.57771E+00	0.16817E+06	0.42854E+00
0.53187E+06	0.69319E+00	0.19753E+06	0.53108E+00
0.56122E+06	0.81401E+00	0.22688E+06	0.63896E+00
0.59058E+06	0.93992E+00	0.25624E+06	0.75193E+00
0.61993E+06	0.10707E+01	0.28559E+06	0.86977E+00
0.64929E+06	0.12061E+01	0.31494E+06	0.99228E+00
0.67864E+06	0.13461E+01	0.34430E+06	0.11193E+01
0.70800E+06	0.14904E+01	0.37365E+06	0.12507E+01
0.73735E+06	0.16389E+01	0.40301E+06	0.13862E+01
0.76671E+06	0.17915E+01	0.43236E+06	0.15259E+01

Table C-7. Continued

TOP LOAD LBS	TOP MOVEMENT INCHES	BASE LOAD LBS	BASE MOVEMENT INCHES
0.79606E+06	0.19481E+01	0.46172E+06	0.16695E+01
0.82541E+06	0.21085E+01	0.49107E+06	0.18170E+01
0.85477E+06	0.22727E+01	0.52042E+06	0.19682E+01
0.88412E+06	0.24405E+01	0.54978E+06	0.21231E+01

NEGATIVE UPWARD DISPLACEMENT

TOP LOAD LBS	TOP MOVEMENT INCHES	BASE LOAD LBS	BASE MOVEMENT INCHES
−0.19877E+05	−0.38437E−02	0.00000E+00	0.00000E+00
−0.44463E+05	−0.18937E−01	0.00000E+00	−0.10000E−01
−0.69052E+05	−0.34038E−01	0.00000E+00	−0.20000E−01
−0.11821E+06	−0.64239E−01	0.00000E+00	−0.40000E−01
−0.21272E+06	−0.12447E+00	0.00000E+00	−0.80000E−01
−0.31375E+06	−0.22746E+00	0.00000E+00	−0.16000E+00
−0.36937E+06	−0.40225E+00	0.00000E+00	−0.32000E+00
−0.36937E+06	−0.72225E+00	0.00000E+00	−0.64000E+00
−0.36937E+06	−0.13623E+01	0.00000E+00	−0.12800E+01
−0.36937E+06	−0.26423E+01	0.00000E+00	−0.25600E+01
−0.36937E+06	−0.52023E+01	0.00000E+00	−0.51200E+01
−0.36937E+06	−0.10322E+02	0.00000E+00	−0.10240E+02
−0.36937E+06	−0.20562E+02	0.00000E+00	−0.20480E+02
−0.36937E+06	−0.41042E+02	0.00000E+00	−0.40960E+02
−0.36937E+06	−0.82002E+02	0.00000E+00	−0.81920E+02
−0.36937E+06	−0.16392E+03	0.00000E+00	−0.16384E+03

STRUC LOAD(LBS)	SOILP(PSF)	ACTIVE DEPTH(FT)
150000.	0.00	50.00

BELL RESTRAINT(LBS)= 449915.44

LOAD-DISPLACEMENT BEHAVIOR

POINT BEARING(LBS)= 37465.96

DEPTH FT	LOAD TRANS LBS	TOTAL LOAD LBS	COM OF INCR INCHES	TOTAL MVMT INCHES	ITER
0.49500E+02	0.35018E+04	0.40968E+05	0.34571E−03	0.82732E−01	2
0.48500E+02	0.35181E+04	0.44486E+05	0.37665E−03	0.83108E−01	2
0.47500E+02	0.35358E+04	0.48022E+05	0.40775E−03	0.83516E−01	2
0.46500E+02	0.35550E+04	0.51577E+05	0.43900E−03	0.83955E−01	2
0.45500E+02	0.35756E+04	0.55152E+05	0.47043E−03	0.84425E−01	2
0.44500E+02	0.35976E+04	0.58750E+05	0.50205E−03	0.84928E−01	2
0.43500E+02	0.36210E+04	0.62371E+05	0.53386E−03	0.85461E−01	2
0.42500E+02	0.36459E+04	0.66017E+05	0.56589E−03	0.86027E−01	2
0.41500E+02	0.36722E+04	0.69689E+05	0.59815E−03	0.86625E−01	2
0.40500E+02	0.37000E+04	0.73389E+05	0.63064E−03	0.87256E−01	2
0.39500E+02	0.32524E+04	0.76641E+05	0.66129E−03	0.87917E−01	2

Table C-7. Continued

DEPTH FT	LOAD TRANS LBS	TOTAL LOAD LBS	COM OF INCR INCHES	TOTAL MVMT INCHES	ITER
0.38500E+02	0.32804E+04	0.79921E+05	0.69008E−03	0.88607E−01	2
0.37500E+02	0.33096E+04	0.83231E+05	0.71913E−03	0.89327E−01	2
0.36500E+02	0.33400E+04	0.86571E+05	0.74844E−03	0.90075E−01	2
0.35500E+02	0.33717E+04	0.89943E+05	0.77802E−03	0.90853E−01	2
0.34500E+02	0.34046E+04	0.93347E+05	0.80789E−03	0.91661E−01	2
0.33500E+02	0.34387E+04	0.96786E+05	0.83805E−03	0.92499E−01	2
0.32500E+02	0.34741E+04	0.10026E+06	0.86852E−03	0.93368E−01	2
0.31500E+02	0.35107E+04	0.10377E+06	0.89931E−03	0.94267E−01	2
0.30500E+02	0.35487E+04	0.10732E+06	0.93042E−03	0.95197E−01	2
0.29500E+02	0.35879E+04	0.11091E+06	0.96188E−03	0.96159E−01	2
0.28500E+02	0.36284E+04	0.11454E+06	0.99369E−03	0.97153E−01	2
0.27500E+02	0.36703E+04	0.11821E+06	0.10259E−02	0.98179E−01	2
0.26500E+02	0.37135E+04	0.12192E+06	0.10584E−02	0.99237E−01	2
0.25500E+02	0.37581E+04	0.12568E+06	0.10913E−02	0.10033E+00	2
0.24500E+02	0.37857E+04	0.12946E+06	0.11246E−02	0.10145E+00	2
0.23500E+02	0.38093E+04	0.13327E+06	0.11581E−02	0.10261E+00	2
0.22500E+02	0.38337E+04	0.13711E+06	0.11918E−02	0.10380E+00	2
0.21500E+02	0.38588E+04	0.14097E+06	0.12257E−02	0.10503E+00	2
0.20500E+02	0.38845E+04	0.14485E+06	0.12598E−02	0.10629E+00	2
0.19500E+02	0.39110E+04	0.14876E+06	0.12941E−02	0.10758E+00	2
0.18500E+02	0.39382E+04	0.15270E+06	0.13287E−02	0.10891E+00	2
0.17500E+02	0.39661E+04	0.15667E+06	0.13636E−02	0.11027E+00	2
0.16500E+02	0.39947E+04	0.16066E+06	0.13987E−02	0.11167E+00	2
0.15500E+02	0.40241E+04	0.16468E+06	0.14340E−02	0.11311E+00	2
0.14500E+02	0.40542E+04	0.16874E+06	0.14696E−02	0.11458E+00	2
0.13500E+02	0.40850E+04	0.17282E+06	0.15055E−02	0.11608E+00	2
0.12500E+02	0.41166E+04	0.17694E+06	0.15417E−02	0.11762E+00	2
0.11500E+02	0.41490E+04	0.18109E+06	0.15781E−02	0.11920E+00	2
0.10500E+02	0.41821E+04	0.18527E+06	0.16148E−02	0.12082E+00	2
0.95000E+01	0.42159E+04	0.18949E+06	0.16518E−02	0.12247E+00	2
0.85000E+01	0.42506E+04	0.19374E+06	0.16891E−02	0.12416E+00	2
0.75000E+01	0.42860E+04	0.19802E+06	0.17268E−02	0.12588E+00	2
0.65000E+01	0.43222E+04	0.20235E+06	0.17647E−02	0.12765E+00	2
0.55000E+01	0.43592E+04	0.20670E+06	0.18030E−02	0.12945E+00	2
0.45000E+01	0.43970E+04	0.21110E+06	0.18416E−02	0.13129E+00	2
0.35000E+01	0.44355E+04	0.21554E+06	0.18805E−02	0.13317E+00	2
0.25000E+01	0.44749E+04	0.22001E+06	0.19198E−02	0.13509E+00	2
0.15000E+01	0.45152E+04	0.22453E+06	0.19594E−02	0.13705E+00	2
0.50000E+00	0.45562E+04	0.22908E+06	0.19994E−02	0.13905E+00	2

INITIAL BASE FORCE(LBS)= 355177.69

ITERATIONS= 81

DEPTH(FT)	LOAD(LBS)	SHAFT MVMT(IN)	SOIL MVMT(IN)
0.00	0.14992E+06	0.98875	2.15238
1.00	0.15721E+06	0.98740	2.58505
2.00	0.16451E+06	0.98598	2.85868
3.00	0.17180E+06	0.98450	3.05836
4.00	0.17909E+06	0.98295	3.20933

Table C-7. Concluded

DEPTH(FT)	LOAD(LBS)	SHAFT MVMT(IN)	SOIL MVMT(IN)
5.00	0.18638E+06	0.98134	3.32392
6.00	0.19367E+06	0.97967	3.40946
7.00	0.20096E+06	0.97793	3.47082
8.00	0.20825E+06	0.97612	3.51146
9.00	0.21554E+06	0.97425	3.53398
10.00	0.22283E+06	0.97232	3.54040
11.00	0.23013E+06	0.97033	3.53233
12.00	0.23742E+06	0.96827	3.51109
13.00	0.24471E+06	0.96614	3.47778
14.00	0.25200E+06	0.96395	3.43333
15.00	0.25929E+06	0.96170	3.37853
16.00	0.26658E+06	0.95938	3.31409
17.00	0.27387E+06	0.95700	3.24058
18.00	0.28116E+06	0.95455	3.15857
19.00	0.28845E+06	0.95204	3.06850
20.00	0.29575E+06	0.94946	2.97082
21.00	0.30304E+06	0.94683	2.86589
22.00	0.31033E+06	0.94412	2.75408
23.00	0.31762E+06	0.94135	2.63568
24.00	0.32491E+06	0.93852	2.51098
25.00	0.33220E+06	0.93563	2.38025
26.00	0.33949E+06	0.93267	2.24373
27.00	0.34678E+06	0.92964	2.10165
28.00	0.35407E+06	0.92655	1.95420
29.00	0.36137E+06	0.92340	1.80157
30.00	0.36866E+06	0.92018	1.64396
31.00	0.37595E+06	0.91690	1.48152
32.00	0.38324E+06	0.91355	1.31441
33.00	0.39019E+06	0.91014	1.14278
34.00	0.39292E+06	0.90669	0.96503
35.00	0.38861E+06	0.90325	0.77876
36.00	0.38207E+06	0.89985	0.58423
37.00	0.37554E+06	0.89651	0.38165
38.00	0.36901E+06	0.89323	0.17124
39.00	0.36248E+06	0.89000	−0.04679
40.00	0.35595E+06	0.88684	−0.27224
41.00	0.34864E+06	0.88373	−0.23257
42.00	0.34133E+06	0.88069	−0.19578
43.00	0.33403E+06	0.87771	−0.16181
44.00	0.32672E+06	0.87480	−0.13064
45.00	0.31941E+06	0.87195	−0.10222
46.00	0.31211E+06	0.86917	−0.07650
47.00	0.30480E+06	0.86645	−0.05346
48.00	0.29749E+06	0.86380	−0.03305
49.00	0.29018E+06	0.86121	−0.01524
50.00	0.28288E+06	0.85868	0.00000

STRUC LOAD(LBS)	SOILP(PSF)	ACTIVE DEPTH(FT)
0.	0.00	−1.00

Table C-8. Listing

```
C         PREDICTION OF VERTICAL SHAFT MOVEMENT
C         DEVELOPED BY LD JOHNSON
          PARAMETER (NL=40,NM=100)
          COMMON AREA,DB,DX,DS,DEL,PI,P(NM),IE(NM),SOILP,NEL,NBX,KOPT
          COMMON/BAS/FORCE,QBULT,CS(NL),CC(NL),EO(NL),PM(NL),AREAB,E50
          COMMON/SHA/ALGTH,AK(NL),B(NL),C(NL),DELT(NM),DELS(NM),
         1   EP(NM),TOT(NM),ZEP(NM),IOPT,JOPT,LOPT,NC,JL,
         2   NCC,FS(11),ZEPP(11),NCURVE(11),ALPHAA(NL)
          DIMENSION TL(NL),TM(NL),R(NL),S(NL),HED(20),T(11,11),Z(11,2),
         1 PS(NL)
          DATA(T(I,1),I=1,11)/.0,.3,.6,.75,.9,.95,1.,1.,1.,1.,1./,
         1   (Z(I,1),I=1,11)/.0,.05,.1,.15,.2,.23,.3,.45,.75,1.05,1.5/
    7     CONTINUE
          Z(1,2)=0.0
          JL=1
          GAW=62.43
          EP(1)=1.E35
          PI=3.1415926
    6     NNN=1
          NC=1
          OPEN(5,FILE='DATLTR.TXT')
          OPEN(6,FILE='LTROUT.TXT')
          OPEN(7,FILE='\GRAPHER\AXIC.DAT')
          OPEN(8,FILE='\GRAPHER\AXIT.DAT')
          OPEN(9,FILE='\GRAPHER\AXIL.DAT')
          READ(5,1) (HED(I),I=1,20)
          WRITE(6,1) (HED(I),I=1,20)
    1     FORMAT(20A4)
          DO 9 I=1,NM
          DELS(I)=0.0
    9     DELT(I)=0.0
          READ(5,2) NMAT,NEL,DX,GWL,LO,IQ,IJ
    2     FORMAT(2I5,2F6.2,4I5)
          WRITE(6,3)NMAT,NEL,DX,GWL,LO,IQ,IJ
    3     FORMAT(//1X,'NMAT=',I5,3X,'NEL=',I5,3X,'DX=',F6.2,' FT',3X,
         1  'GWL= ',F6.2,' FT',/,3X,'LO=',I5,3X,'IQ (SHAFT INC)=',I5,
         2  3X,'IJ (NO.LOADS)=',I5)
          LOPT=LO
          NNP=NEL+1
          READ(5,4)IOPT,JOPT,KOPT,SOILP,DS,DB
    4     FORMAT(3I5,3F10.3)
          WRITE(6,5)IOPT,JOPT,KOPT,SOILP,DS,DB
    5     FORMAT(//1X,'I=',I5,3X,'J=',I5,3X,'K=',I5,10X,'SOILP=',F10.2,
         1   ' PSF',/,1X,'DS=',F10.2,' FT',/,1X,'DB=',F10.2,' FT')
          IF(KOPT.EQ.2.OR.KOPT.EQ.5.OR.KOPT.EQ.9)READ(5,11)E50
   11     FORMAT(E13.3)
          IF(KOPT.EQ.2.OR.KOPT.EQ.5.OR.KOPT.EQ.9)WRITE(6,12)E50
   12     FORMAT(//1X,'E50=',E13.3)
          IU=IJ
          READ(5,13)LLL
   13     FORMAT(I5)
          IF(LLL.EQ.0)WRITE(6,14)LLL
```

Table C-8. Continued

```
 14     FORMAT(//1X,'LOCAL SHEAR FAILURE AT BASE - LLL=',I5)
        IF(LLL.EQ.1)WRITE(6,15)LLL
 15     FORMAT(//1X,'GENERAL SHEAR FAILURE AT BASE - LLL=',I5)
        WRITE(6,10)
 10     FORMAT(//1X,'MAT  GS  EO  WO(%)  PS(PSF)  CS  CC  CO(PSF)',
       1  ' PHI  K  PM(PSF)',/)
  8     READ(5,16)M,TL(M),EO(M),TM(M),PS(M),CS(M),CC(M),C(M),B(M),AK(M),
       1  PM(M)
 16     FORMAT(I3,3F6.2,F7.0,2F7.2,F7.0,2F6.2,F7.0)
        WRITE(6,16)M,TL(M),EO(M),TM(M),PS(M),CS(M),CC(M),C(M),B(M),AK(M),
       1  PM(M)
        B(M)=B(M)*PI/180.
        IF(PM(M).LT.PS(M)) PM(M) = PS(M)
        IF(NMAT-M)18,20,8
 18     PRINT 17,M
 17     FORMAT(17HERROR IN MATERIAL,I5)
        GO TO 8
 20     IF(IOPT.NE.6)GO TO 23
        READ(5,21)(ALPHAA(I),I=1,NMAT)
 21     FORMAT(2(7F10.5))
        WRITE(6,22)(ALPHAA(I),I=1,NMAT)
 22     FORMAT(//1X,'ALPHA=',2(7F10.5))
 23     L=0
        WRITE(6,25)
 25     FORMAT(//1X,'ELEMENT',5X,'NO OF SOIL',/)
 30     READ(5,26)M,IE(M)
 26     FORMAT(2I5)
 35     L=L+1
        IF(M-L)45,45,40
 40     IE(L)=IE(L-1)
        GO TO 35
 45     IF(NEL-L)50,50,30
 50     CONTINUE
        DO 52 I=1,NEL
        WRITE(6,53)I,IE(I)
 52     CONTINUE
 53     FORMAT(1X,I5,10X,I5)
C       EFFECTIVE STRESS IN SOIL
        P(1)=0.0
        DXX=DX
        DO 55 I=2,NNP
        MTYP=IE(I-1)
        WC=TM(MTYP)/100.
        GAM=TL(MTYP)*GAW*(1.+WC)/(1.+EO(MTYP))
        IF(DXX.GT.GWL)GAM=GAM-GAW
        P(I)=P(I-1)+DX*GAM
        DXX=DXX+DX
 55     CONTINUE
        IF(KOPT.LE.6)GO TO 56
        READ(5,57)RFF,GG
 57     FORMAT(F6.3,E13.3)
        WRITE(6,59)RFF,GG
```

Table C-8. Continued

```
59      FORMAT(//1X,'REDUCTION FACTOR=',F6.3,3X,'SHEAR MODULUS=',E13.3)
        HALF=FLOAT(IQ)/2.
        IJK=IFIX(HALF)
        MTYP=IE(IJK)
        PH=B(MTYP)
        SHR=C(MTYP)+AK(MTYP)*SIN(PH)*P(IJK)/COS(PH)
        CONS=SHR*DS/(2.*GG)
        D1=1.5*FLOAT(IQ)*DX/DS
        DO 58 I=1,11
        CONC=(D1-T(I,1)*RFF)/(1.-T(I,1)*RFF)
        Z(I,1)=12.*T(I,1)*CONS*ALOG(CONC)
58      CONTINUE
C       LOAD-DEFLECTION BEHAVIOR OF SHAFT
56      IF(KOPT.LE.2.OR.KOPT.GT.6)GO TO 72
        READ(5,62)NCA
62      FORMAT(I5)
        WRITE(6,60)NCA
60      FORMAT(//1X,'NO. OF LOAD-TRANSFER CURVES(<12)?=',I5)
        DO 65 I=1,NCA
        WRITE(6,61)I
61      FORMAT(/1X,'CURVE ',I5,/)
        II=I+1
        READ(5,63)(T(M,II),M=1,11)
63      FORMAT(11F6.2)
        WRITE(6,64)(T(M,II),M=1,11)
64      FORMAT(1X,'RATIO SHR DEV,M=1,11 ARE',/1X,11F6.3)
65      CONTINUE
        DO 67 M=2,11
        READ(5,66)S(M)
        IF(S(M).LT..000001)GO TO 68 JL=2
        Z(M,2)=S(M)
        WRITE(6,69)M,S(M)
67      CONTINUE
68      CONTINUE
66      FORMAT(F6.3)
69      FORMAT(/1X,'MOVEMENT(IN.) FOR LOAD TRANSFER M=',I5,' IS ',F6.3,
       1    '      INCHES')
72      IF(IOPT.LE.4.OR.IOPT.EQ.6)GO TO 78
        READ(5,81)NCC
81      FORMAT(I5)
        WRITE(6,82)NCC
82      FORMAT(//1X,'NO OF SKIN FRICTION-DEPTH TERMS(<12)? ARE',I5,/1X,
       1    'SKIN FRICTION(PSF) DEPTH(FT) CURVE NO',//)
        DO 89 M=1,NCC
        READ(5,83)FS(M),ZEPP(M),NCUR
        WRITE(6,77)FS(M),ZEPP(M),NCUR
77      FORMAT(3X,F10.3,5X,F10.3,5X,I5)
83      FORMAT(2F10.3,I5)
        NCURVE(M)=1+NCUR
89      CONTINUE
78      IF(JOPT.EQ.1)GO TO 84
        READ(5,81)NC
        READ(5,92)(EP(M),ZEP(M),M=1,NC)
```

Table C-8. Continued

```
 92     FORMAT(E13.3,F6.2)
        WRITE(6,102)(EP(M),ZEP(M),M=1,NC)
102     FORMAT(//1X,'E SHAFT(PSF) AND DEPTH(FT):',/1X,4(E13.3,2X,F6.2),
       1  /1X,4(E13.3,2X,F6.2),/1X,4(E13.3,2X,F6.2),/1X,4(E13.3,2X,F6.2))
 84     UU=100.
        MM=0
        ALGTH=DX*FLOAT(IQ)
        AREA=(PI/4.)*DS**2.
        AREAB=(PI/4.)*DB**2.
        NBX=IQ+1
        IF(NBX.LT.NNP)MAT=IE(NBX)
        IF(NBX.EQ.NNP)MAT=IE(NEL)
        QBU=9.*C(MAT)
        IF(LLL.LE.0.1)QBU=7.*C(MAT)
        IF(B(MAT).LT.0.001)GO TO 86
        PH=B(MAT)
        TA=SIN(PH)/COS(PH)
        XNQ=(PH/2.)+45.*PI/180.
        IF(LLL.EQ.0)GO TO 74
        TB=(1.5*PI-PH)*TA
        XNQ=EXP(TB)/(2.*(COS(XNQ))**2.)
        GO TO 76
 74     XNQ=(1.+TA)*EXP(TA)*(SIN(XNQ)/COS(XNQ))**2. 76 XNC=(XNQ-1.)/TA
        QBU=C(MAT)*XNC+(SOILP+P(NBX))*XNQ
 86     QBULT=QBU*AREAB
        NN=1
        NO=IU
        ITER=1
        S(1)=(SOILP+P(NBX))*AREAB
        IF(KOPT.NE.6)GO TO 91
        R(1)=0.0
        IU1=IU-1
        WRITE(6,95)S(1),IU1
 95     FORMAT(//,1X,'BASE DISPLACEMENT(IN.),BASE LOAD(LBS) >',F10.2,
       1      ' FOR',I5,' POINTS')
        DO 94 M=2,IU
        READ(5,103)R(M),S(M)
        WRITE(6,103)R(M),S(M)
 94     CONTINUE
103     FORMAT(F10.5,F15.3)
 91     WRITE(6,90)QBULT
 90     FORMAT(/,10X,'BEARING CAPACITY=',F13.2,' POUNDS')
        WRITE(6,88)
 88     FORMAT(/,1X,'DOWNWARD DISPLACEMENT')
        IF(KOPT.NE.6)DLOAD=(QBULT-S(1))/(FLOAT(IU)-1.)
 87     IF(ITER.GT.1.AND.KOPT.NE.6)S(ITER)=S(ITER-1)+DLOAD
        FORCE=S(ITER)
        IF(KOPT.NE.6)CALL BASEL
        IF(KOPT.NE.6)R(ITER)=DEL
        IF(KOPT.EQ.6)DEL=R(ITER)
 98     TOT(NBX)=FORCE
        GO TO 97
 93     WRITE(6,99)
```

Table C-8. Continued

```
 99    FORMAT(/,1X,'NEGATIVE UPWARD DISPLACEMENT')
 96    R(ITER)=DEL
       TOT(NBX)=0.0
       S(ITER)=0.0
 97    CALL SHAFL(T,Z)
       TL(ITER)=TOT(1)
       TM(ITER)=DELT(1)
       IF(ITER-NO)100,101,101
100    ITER=ITER+1
       IF(MM.EQ.0)GO TO 87
       XN=FLOAT(ITER-IU-2)
       DEL=-0.01*2.**XN
       GO TO 96
101    WRITE(6,105)
105    FORMAT(/,49H TOP LOAD TOP MOVEMENT BASE LOAD BASE MOVE,
      1     4HMENT,/,49H   LBS        INCHES           LBS         INCHES)
       DO 110 M=NN,NO
       WRITE(6,111)TL(M),TM(M),S(M),R(M)
       IF(MM.EQ.0) WRITE(7,111)TL(M),-TM(M),S(M),-R(M)
       IF(MM.EQ.1) WRITE(8,111)TL(M),-TM(M),S(M),-R(M)
110    CONTINUE
111    FORMAT(4E13.5)
112    MM=MM+1
       DEL=0.0
       IF(MM.EQ.2)GO TO 125
       ITER=IU+1
       NO=2*IU
       NN=IU+1
       GO TO 93
C      SHAFT MOVEMENT FOR STRUCTURAL LOAD
125    WRITE(6,121)
121    FORMAT(/,1X,'STRUC LOAD(LBS)   SOILP(PSF)   ACTIVE DEPTH(FT)')
       READ(5,122)STRUL,SOILP,XA
       WRITE(6,1220)STRUL,SOILP,XA
122    FORMAT(3F15.2)
1220   FORMAT(3X,F10.0,3X,F10.2,5X,F10.2)
       IF(XA.LT.0.0)GO TO 380
       RESTU=0.0
       DD=1000.
       DEL=0.0
       DELL=0.0
       LOPT=0
       IF(LO.EQ.3)LOPT=3
       V=0.0025
       DO 123 I=1,NNP
123    DELS(I)=0.0
       ITER=1
       STRU=STRUL
       QBR=(7./9.)*QBU
124    RESTR=(QBR+SOILP+P(NBX-1))*PI*(DB**2.-DS**2.)/4.
       IF(DB.LE.DS)RESTR=0.0
       IF(STRU.LT.0.0)RESTU=RESTR+(QBR+ALGTH
      1  *150.)*AREA
       WRITE(6,126)RESTR
```

Table C-8. Continued

```
      126   FORMAT(/,1X,'BELL RESTRAINT(LBS)=',8X,F13.2)
            IF(STRU.LT.0.0)WRITE (6,127) RESTU
      127   FORMAT(/,1X,'FIRST ESTIMATE OF PULLOUT RESTRAINT(LBS)=',F13.2)
            NN=1
            NO=IU
            WRITE(6,128)
      128   FORMAT(/,20X,26HLOAD-DISPLACEMENT BEHAVIOR,/)
            IF(STRU.GT.0.0)GO TO 135
            FORCE=STRU EXFOR=RESTR+FORCE
            IF(EXFOR.GT.-0.01)DEL=0.0
            IF(EXFOR.GT.-0.01)GO TO 145
            STRU=EXFOR
            NN=IU+1
            NO=2*IU
      135   DO 136 M=NN,NO
            IF(M.GT.IU)GO TO 141
            IF(TL(M)-STRU)136,137,137
      141   IF(ABS(TL(M))-ABS(STRU))136,137,137
      136   CONTINUE
            DIFF=TL(M)-TL(M-1)
            IF(DIFF.LT.0.01)DEL=R(M)
            IF(DIFF.LT.0.01)GO TO 138
      137   DEL=R(M-1)+(R(M)-R(M-1))*(STRU-TL(M-1))/(TL(M)-TL(M-1))
      138   IF(ABS(STRU).GT.ABS(TL(M)).AND.STRU.LT.0.0)GO TO 376
            TOT(NBX)=S(M-1)+(S(M)-S(M-1))*(STRU-TL(M-1))/(TL(M)-TL(M-1))
            CALL SHAFL(T,Z)
            DEL=DELT(NBX)
C           VERTICAL DISPLACEMENT OF ADJACENT SOIL
      145   IF(LO.LT.2)WRITE(6,146)
      146   FORMAT(//20X,'EFFECT OF ADJACENT SOIL'/)
            IF(LO.GE.2)LOPT=3
            DO 150 K=2,NNP
            I=NNP+1-K
            DEPTH=DX*FLOAT(I)-DX/2.
            IF(DEPTH.GT.XA)GO TO 150
            MTYP=IE(I)
            TEM=(SOILP+(P(I)+P(I+1))/2.)/PS(MTYP)
            TEA = (SOILP+(P(I)+P(I+1))/2.)/PM(MTYP)
            TEB = PM(MTYP)/PS(MTYP)
            TEC = DX*12./(1.+EO(MTYP))
            IF(TEA.LE.1.)TDEL=CS(MTYP)*TEC*ALOG10(TEM)
            IF(TEA.GT.1.)TDEL=TEC*(CS(MTYP)*ALOG10(TEB)+CC(MTYP)*ALOG10(TEA))
            DELS(I)=TDEL+DELS(I+1)
      150   CONTINUE
C           LOAD TRANSFER TO SHAFT FROM SOIL MOVEMENT
            ITT=1
            IF(NNN.GT.1)LOPT=1
      160   TOT(NBX)=0.0
            CALL SHAFL(T,Z)
            FORCE=TOT(NBX)-TOT(1)+STRUL
            IF(ITER.EQ.1)WRITE(6,162)FORCE
      162   FORMAT(/,1X,'INITIAL BASE FORCE(LBS)=',F13.2,/)
            IF(FORCE.LT.0.0.OR.ITER.GT.1)GO TO 300
```

Table C-8. Continued

```
C         DOWN FORCE, POSITIVE TIP DISPLACEMENT
          BBETA=0.0
    170   IF(KOPT.NE.6)CALL BASEL
C         SHAFT MOVEMENT FOR DOWNDRAG
          IF(KOPT.NE.6)GO TO 200
          DO 163 M=2,IU
          IF(S(M)-FORCE)163,164,164
    163   CONTINUE
    164   DEL=R(M-1)+(R(M)-R(M-1))*(FORCE-S(M-1))/(S(M)-S(M-1))
    200   TOT(NBX)=FORCE
          LOPT=LO
          CALL SHAFL(T,Z)
          BETA=TOT(1)-STRUL
          IF(LO.LT.2)WRITE(6,309)ITER,FORCE,BETA
          IF(UU-ABS(BETA))205,350,350
    205   IF(BETA.LT.0.0)GO TO 206
          IF(BBETA.LT.-50..AND.DD.LT.101.)GO TO 350
          IF(BBETA.LT.-50..AND.DD.GT.999.)DD=100.
          BBETA=BETA
          FORCE=TOT(NBX)-DD
          IF(FORCE.LT.0.0)ITER=1
          IF(FORCE.LT.0.0)GO TO 300
          ITER=ITER+1
          GO TO 170
    206   IF(BBETA.GT.50..AND.DD.LT.101.)GO TO 350
          IF(BBETA.GT.50..AND.DD.GT.999.)DD=100.
          BBETA=BETA
          FORCE=TOT(NBX)+DD
          ITER=ITER+1
          GO TO 170
C         UPLIFT FORCE, NEGATIVE TIP DISPLACEMENT
    300   LOPT=LO
          IF(ITER.EQ.1)DELL=DELT(NBX)
          IF(LOPT.GE.1)GO TO 307
    302   WRITE(6,305)FORCE
    305   FORMAT(/,1X,'BASE FORCE(LBS)=',3X,F13.2)
    307   EXFOR=RESTR+FORCE
          IF(EXFOR.GT.0.0)GO TO 340
          IF(ITT.EQ.100)WRITE(6,311)DELL,FORCE
          IF(ITT.EQ.100)ITT=0
          DELL=DELL-V
          DEL=DELL
          IF(DEL.LT.DELS(1))GO TO 375
          LOPT=LO
          ITT=ITT+1
          ITER=ITER+1
          GO TO 160
    309   FORMAT(25X,I5,2F15.2)
    311   FORMAT(/,1X,'DISPLACEMENT(INCHES)=',3X,F8.4,4X,'FORCE=',F12.2,
         1  '         POUNDS')
C         SHAFT MOVEMENT FOR UPLIFT
```

Table C-8. Continued

```
 340     TOT(NBX)=FORCE
         PRINT 311,DELL,FORCE
         DEL=DELL
         LOPT=LO
         IF(LO.EQ.1)LOPT=0
         CALL SHAFL(T,Z)
 350     FORCE=TOT(NBX)-TOT(1)+STRUL
         WRITE(6,352)ITER
 352     FORMAT(/,1X,'ITERATIONS=',I10)
         IF(FORCE.GT.QBULT)WRITE(6,353)
 353     FORMAT(/,1X,'THE SOIL BEARING CAPACITY IS EXCEEDED')
         WRITE(6,355)
 355     FORMAT(/,1X,'DEPTH(FT) LOAD(LBS) SHAFT MVMT(IN)',
        1     '    SOIL MVMT(IN)')
         DXX=0.0
         DO 360 I=1,NBX
         TOTAL=DELT(I)+DELS(I)
         IF(FORCE.GT.0.0)TOTAL=TOTAL-DELS(NBX)
         WRITE(6,370)DXX,TOT(I),TOTAL,DELS(I)
         WRITE(9,370)-DXX,TOT(I),TOTAL,DELS(I)
         DXX=DXX+DX
 360     CONTINUE
 370     FORMAT(F7.2,2X,E13.5,2F15.5)
         GO TO 379
 376     WRITE(6,377)
 377     FORMAT(/,1X,'SHAFT PULLS OUT')
         GO TO 379
 375     WRITE(6,378)
 378     FORMAT(//,1X,'SHAFT UNSTABLE')
         WRITE(6,311)DELL,FORCE
 379     NNN=NNN+1
         GO TO 125
 380     READ(5,81)NON
         IF(NON.GE.1)GO TO 7
         CLOSE(5,STATUS='KEEP')
         CLOSE(6,STATUS='KEEP')
         CLOSE(7,STATUS='KEEP')
         CLOSE(8,STATUS='KEEP')
         CLOSE(9,STATUS='KEEP')
         STOP
         END
C
C
         SUBROUTINE BASEL
         PARAMETER (NL=40,NM=100) COMMON AREA,DB,DX,DS,DEL,PI,P(NM),IE(NM),SOILP,NEL,NBX,KOPT
         COMMON AREA, DB, DX, DS, DEL, PI, P(NM), IE(NM), SOILP, NEL, NBX, KOPT
         COMMON/BAS/FORCE,QBULT,CS(NL),CC(NL),EO(NL),PM(NL),AREAB,E50
         DIMENSION Q(NM)
         DXX=DX
         NNP=NEL+1
         NA=NBX+1
         BPRES=FORCE+(DB**2.-DS**2.)*(SOILP+P(NBX))*PI/4.
         IF(DB.LT.DS)BPRES=FORCE
```

Table C-8. Continued

```
      BPRES=BPRES/AREAB
      IF(KOPT.EQ.1.OR.KOPT.EQ.4.OR.KOPT.EQ.8)GO TO 15
      IF(KOPT.EQ.2.OR.KOPT.EQ.5.OR.KOPT.EQ.9)GO TO 14
      Q(NBX)=BPRES
      DO 5 I=NA,NNP
      PT=1.+(DB/(2.*DXX))**2.
      PT=PT**1.5
      Q(I)=P(I)+SOILP+(BPRES-P(NBX)-SOILP)*(1.-1./PT)
      DXX=DXX+DX
 5    CONTINUE
      DEL=0.0
      IF(NBX.GT.NEL)GO TO 20
      DO 10 I=NBX,NEL
      MTYP=IE(I)
      QQ=(Q(I)+Q(I+1))/2.
      PPP=SOILP+(P(I)+P(I+1))/2.
      F=QQ/PPP
      CON=CC(MTYP)
      IF(QQ.LT.PM(MTYP))CON=CS(MTYP)
      IF(QQ.GT.PM(MTYP))GO TO 8
      DEL=DEL+DX*12.*CON*ALOG10(F)/(1.+EO(MTYP))
      GO TO 10
 8    AA=PM(MTYP)/PPP
      AB=QQ/PM(MTYP)
      DEL=DEL+(CS(MTYP)*ALOG10(AA)+CC(MTYP)*ALOG10(AB))*DX*12./
     1  (1.+EO(MTYP))
 10   CONTINUE
      GO TO 20
 14   C1=0.76
      C2=1.5
      C3=2.*E50
      GO TO 16
 15   C1=1.
      C2=3.
      C3=0.04
 16   DEL=((((BPRES)*AREAB)/(QBULT*C1))**C2)*C3*DB*12.
      DEL=DEL-((((P(NBX)+SOILP)*AREAB)/(QBULT*C1))**C2)*C3*DB*12.
 20   CONTINUE
      RETURN END
C
C
      SUBROUTINE SHAFL(T,Z)
      PARAMETER (NL=40,NM=100)
      COMMON AREA,DB,DX,DS,DEL,PI,P(NM),IE(NM),SOILP,NEL,NBX,KOPT
      COMMON/SHA/ALGTH,AK(NL),B(NL),C(NL),DELT(NM),DELS(NM),
     1 EP(NM),TOT(NM),ZEP(NM),IOPT,JOPT,LOPT,NC,JL,
     1 NCC,FS(11),ZEPP(11),NCURVE(11),ALPHAA(NL)
      DIMENSION T(11,11),Z(11,2)
      U=0.0001
      DELC=0.0
      IF(JOPT.EQ.0)GO TO 5
      EPILE=EP(1)
```

Table C-8. Continued

```
  5     IF(LOPT.GE.1)GO TO 16
        WRITE(6,10)TOT(NBX)
 10     FORMAT(19HPOINT BEARING(LBS)=,F13.2,//)
        WRITE(6,15)
 15     FORMAT(50H   DEPTH   LOAD TRANS   TOTAL LOAD   COM OF IN,
       1  20HCR   TOTAL MVMT   ITER,/,31H   FT   LBS ,
       2  31H LBS   INCHES   INCHES)
 16     DELT(NBX)=DEL
        DO 50 K=2,NBX
        I=NBX+1-K
        MTYP=IE(I)
        NTER=1
        AKK=K-2
        BSLMT=ALGTH-(AKK*DX)
        DEPTH=BSLMT-DX/2.
        IF(IOPT.LE.4.OR.IOPT.EQ.6)GO TO 17
        DO 18 JJ=2,NCC
        IF(ZEPP(JJ)-DEPTH)18,19,19
 18     CONTINUE
 19     JK=NCURVE(JJ)
        SKF=FS(JJ-1)+(FS(JJ)-FS(JJ-1))*(DEPTH-ZEPP(JJ-1))/(ZEPP(JJ)-ZE
       1    PP(JJ-1))
 17     IF(IOPT.LE.4.OR.IOPT.EQ.6)JK=1
        DO 20 M=2,11
        IF(Z(M,JL)-ABS(DEL))20,21,21
 20     CONTINUE
        PERCP=T(11,JK)
        GO TO 25
 21     PERCP=T(M-1,JK)+((T(M,JK)-T(M-1,JK))*(ABS(DEL)-Z(M-1,JL))/
       1  (Z(M,JL)-Z(M-1,JL)))
 25     IF(IOPT.EQ.5)SHR=SKF*PERCP
        IF(IOPT.EQ.5)GO TO 35
 26     F=BSLMT*PI/ALGTH PH=B(MTYP)
        TA=SIN(PH)/COS(PH)
        SHR=TA*AK(MTYP)*(SOILP+(P(I)+P(I+1))/2.)+C(MTYP)
        IF(IOPT.EQ.1)SHR=SHR*SIN(F)
        IF(IOPT.EQ.2)SHR=SHR*0.6
        IF(IOPT.EQ.3)SHR=SHR*.45
        IF(IOPT.EQ.4)SHR=SHR*.3
        IF(IOPT.EQ.6)SHR=SHR*ALPHAA(MTYP)
 35     TNSLD=SHR*PERCP
        IF(DEL.LT.0.0)TNSLD=-TNSLD
        ALDINC=TNSLD*PI*DS*DX-(P(I+1)-P(I))*AREA
        IF(JOPT.EQ.1.OR.NC.EQ.1)GO TO 42
        DO 40 M=2,NC
        IF(ZEP(M)-DEPTH)40,41,41
 40     CONTINUE
 41     EPILE=EP(M-1)+(EP(M)-EP(M-1))*(DEPTH-ZEP(M-1))/(ZEP(M)-
       1  ZEP(M-1))
```

Table C-8. Concluded

```
42    TOT(I)=ALDINC+TOT(I+1)
      AVGLD=TOT(I+1)+ALDINC/2.
      IF(JOPT.EQ.1)GO TO 46
      AEPILE=AREA*EPILE
      DELC=AVGLD*DX*12./AEPILE
      DELT(I)=DELC+DELT(I+1)-(DELS(I)-DELS(I+1))
      DELAV=((((TOT(I)+TOT(I+1))/2.)+TOT(I+1))*DX*3./AEPILE)
     1  +DELT(I+1)
      ETA=DELAV-DEL
      IF(U-ABS(ETA))45,46,46
45    DEL=DELAV
      NTER=NTER+1
      GO TO 17
46    IF(LOPT.GE.1)GO TO 48
      WRITE(6,55)DEPTH,ALDINC,TOT(I),DELC,DELT(I),NTER
48    DEL=DELT(I)
50    CONTINUE
55    FORMAT(5E13.5,I5)
      RETURN
      END
```

APPENDIX D

NOTATION

Symbol	Description
c	Unit soil cohesion, ksf; distance from centroid to outer fiber, ft
c_{st}	Distance from centroid of steel reinforcing rod to outer fiber, ft
c'	Effective unit soil cohesion, kips per square foot (ksf)
c_a	Adhesion of soil to base ≤ c, ksf
d_r	Diameter of vane rod, inch
d_v	Vane diameter, inch
e	Void ratio
e_{max}	Reference void ratio of a soil at the minimum density
e_{min}	Reference void ratio of a soil at the maximum density
e_B	Eccentricity parallel with B, M_B/Q, ft
e_W	Eccentricity parallel with W, M_W/Q, ft
f_n	Negative skin friction, ksf
f_{ni}	Mobilized negative skin friction of pile element i, ksf
f_s	Skin friction, ksf
f_{si}	Skin friction of pile element i, ksf
f'_c	Concrete strength, psi
f'_{ys}	Steel yield strength, psi
$\bar{f}_s$	Full mobilized skin friction, ksf
h	Height of hammer fall, ft
h_v	Vane height, inch
k	Constant relating elastic soil modulus with depth $E_s = kz$, kips/ft^3
	Term preventing unlimited increase in bearing capacity with increasing depth for Hanson method
k_c	Point correlation factor used in CPT B & G method
n	Number of piles in a group, number of pile elements
p_o	Internal pressure causing lift-off of dilatometer membrane, ksf
p_1	Internal pressure required to expand central point of the dilatometer membrane by 1.1 millimeters, ksf
p_L	Pressuremeter limit pressure, ksf
q	Bearing pressure on foundation, ksf
q_1	Soil pressure per inch of settlement, ksf
q_a	Allowable unit bearing capacity, ksf
q_b	Unit base resistance, ksf
q_{bu}	Unit ultimate end bearing resistance, ksf
q_c	Cone penetration resistance, ksf
q_{c1}	Average q_c over a distance of $L + 0.7B$ to $L + 4B$ below pile tip, Figure 5-21, ksf
q_{c2}	Average q_c over a distance L to $L - 8B$ above pile tip, Figure 5-21, ksf
q_{cb1}	Average cone penetration resistance from footing base to 0.5B below base, ksf
q_{cb2}	Average cone penetration resistance from 0.5B to 1.5B below base, ksf
q_{ci}	Cone penetration resistance of depth increment i, ksf
$\bar{q}_c$	Equivalent cone penetration resistance from footing base to 1.5B below base, ksf
q_d	Design unit bearing pressure, ksf
q_{load}	Area pressure applied to soil supporting pile, ksf
q_{na}	Nominal unit allowable bearing capacity, ksf
q_r	Resultant applied pressure on foundation soil, R/BW, ksf

Symbol	Description
q_u	Ultimate unit bearing capacity, ksf
q_{ua}	Ultimate unit bearing capacity of axially loaded square or round footings with horizontal ground surface and base, kips
q_{ut}	Ultimate unit bearing capacity of upper dense sand, ksf
$q_{a,1}$	Allowable unit bearing capacity for 1 inch of settlement, ksf
$q_{u,b}$	Ultimate unit bearing capacity on a very thick bed of the bottom soft clay layer, ksf
$q_{u,p}$	Ultimate unit bearing capacity of plate, ksf
q_ℓ	Limiting stress for Meyerhof method $N_{qp}\tan\phi'$, ksf
q'_u	Net ultimate bearing capacity, $q_u - \gamma_D \cdot D$, ksf
r_γ	Reduction factor, $1 - 0.25\log(B/6)$
s	Spacing between piles, ft
u_w	Pore water pressure, ksf
y_a	Allowable lateral deflection, inch
y_o	Lateral groundline deflection, inch
z	Depth, ft
A	Cross-section area of drilled shaft or pile, ft^2
A_b	Area of tip or base, ft^2
A_{bp}	Area of base resisting pullout force, ft^2
A_e	Effective area of foundation $B'W'$, ft^2
A_{si}	Perimeter area of pile element i, $C_{si} \cdot \Delta L$
A_{st}	Area of steel, inch^2
B	Least lateral dimension of a foundation or pile diameter, ft
B_b	Base diameter, ft
B_{dia}	Diameter of circular footing, ft
B_p	Diameter or width of the plate, ft
B_r	Horizontal distance beneath center of strip footing to location of outermost rod in reinforced soil, ft
B_s	Diameter or width of pile or shaft, ft
B'	Effective foundation width, $B - 2e_B$, ft
C_f	Correction factor for K when $\delta \neq \phi'$
C_g	Circumference of pile group, minimum length of line that can enclose pile group, ft
C_{ua}	Average undrained shear strength of cohesive soil in which the group is placed, ksf
C_{ub}	Average undrained shear strength of cohesive soil below the tip to a depth $2B_b$ below the tip, ksf
C_{um}	Mean undrained shear strength along pile length, ksf
C_{ov}	Overburden pressure adjustment $(\sigma_o/\sigma'_v)^{0.5}$
C_s	Circumference of drilled shaft or pile, ft
C_{si}	Circumference of drilled shaft or pile element i, ft
C_u	Undrained cohesion, ksf
$C_{u,lower}$	Undrained shear strength of the soft lower clay, ksf
$C_{u,upper}$	Undrained shear strength of the stiff upper clay, ksf
C_z	Pile Perimeter at depth z, ft
C_{ER}	Rod energy correction factor
C_L	Perimeter of the pile tip, ft
C_N	Overburden correction factor
CPT	Cone penetration test
D	Depth of the foundation base below ground surface, ft
D_c	Critical depth where increase in stress from structure is 10 percent of the vertical soil stress beneath the foundation, ft
D_e	Equivalent embedment depth using CPT procedure for estimating bearing capacity, ft
D_r	Relative density, percent
D_R	Relative density, fraction
D_{GWT}	Depth below ground surface to groundwater, ft

Symbol	Description
E_g	Efficiency of pile group
E_h	Hammer efficiency
E_p	Young's modulus of pile, ksf (kips/inch^2)
E_s	Elastic soil modulus, ksf
E_{sl}	Lateral modulus of soil subgrade reaction, ksf
F_r	Reduction factor for drilled shaft unit end bearing capacity
FS	Factor of safety
G	Specific gravity
G_i	Initial shear modulus, ksf
G_s	Shear modulus, ksf
H	Depth of shear zone beneath base of foundation, ft
H_b	Vertical distance from the shaft base in a group to the top of the weak layer, ft
H_r	Height of vertical reinforcement rods placed in soil supporting a strip foundation, ft
H_t	Depth below footing base to weak stratum or soft clay, ft
I_c	Moment of inertia of concrete section, ft^4
I_p	Moment of inertia of pile, ft^4
I_r	Rigidity index
I_{rr}	Reduced rigidity index
I_{st}	Moment of inertia of steel section, ft^4
I_D	Material deposit index of dilatometer test
K	Lateral earth pressure coefficient
K_o	Coefficient of earth pressure at rest
K_p	Rankine coefficient of passive pressure, $\tan^2\left(45 + \frac{\phi}{2}\right)$ or $\frac{1 + \sin\phi}{1 - \sin\phi}$
K_{ps}	Punching shear coefficient
K_v	Constant depending on dimensions and shape of the vane, ft^3
K_D	Horizontal stress index of dilatometer test
L	Embeded length of deep foundation, ft
Lc	Critical depth at which increasing pile lengths provide no increase in end bearing resistance for Meyerhof's method, ft
L_c	Critical length between long and short pile, ft
L_{cs}	Critical length between short and intermediate pile, ft
L_{cl}	Critical length between intermediate and long pile, ft
L_{clay}	Length of pile in clay, ft
L_n	Length to neutral point *n*, ft
L_{sand}	Length of pile in sand, ft
L_{sh}	Horizontal length of shear zone at the foundation depth, ft
M_a	Applied bending moment on pile butt (top) in clockwise direction, kips-ft
M_y	Ultimate resisting bending moment of entire pile cross-section, kips-ft
M_B	Bending moment parallel with *B*, kips-ft
M_W	Bending moment parallel with *W*, kips-ft
N_c	Dimensionless bearing capacity related with cohesion
N_{cp}	Pile dimensionless bearing capacity related with cohesion
N_k	Cone factor relating undrained cohesion with cone penetration resistance, often varies from 14 to 20
N_n	Standard penetration resistance correlated to n percent energy, blows/foot
N_p	Relationship between shear modulus and undrained cohesion used in pressuremeter test, $1 + \ln(G_s/C_u)$
N_q	Dimensionless bearing capacity factor related with surcharge
N_{qp}	Pile dimensionless bearing capacity factor related with surcharge
N_{SPT}	Average blow per foot in the soil produced by a 140 pound hammer falling 30 inches to drive a standard sampler (1.42′ I.D., 2.00″ O.D.) one foot
N_{60}	Penetration resistance normalized to an effective energy delivered to the drill rod at 60 percent of theoretical free-fall energy, blows/foot

Symbol	Description
N_{70}	Penetration resistance normalized to an effective energy delivered to the drill rod at 70 percent of theoretical free-fall energy, blows/foot
N_γ	Dimensionless bearing capacity factor related with soil weight in the failure wedge
$N_{\gamma p}$	Pile dimensionless bearing capacity factor related with soil weight in the failure wedge
N_ϕ	$\tan^2\left[45 + \frac{\phi}{2}\right]$
OCR	Overconsolidation ratio
P	Pullout load, kips
PI	Plasticity index, percent
P_{max}	Maximum tensile force in shaft, kips
P_{nu}	Pullout skin resistance force, kips
P_{nui}	Pullout skin resistance for pile element i, kips
P_u	Ultimate pullout resistance, kips
Q	Vertical load on foundation, kips
Q_a	Allowable bearing capacity force, kips
Q_b	Base resistance force, kips
Q_{bu}	Base resistance capacity, kips
Q_{bur}	Ultimate base resistance of upper portion of underream, kips (pounds)
Q_d	Design bearing force, kips
Q_e	Applied load in elastic range, kips
Q_s	Soil-shaft side friction resistance, kips
Q_{su}	Soil-shaft side friction resistance capacity or uplift thrust, kips
Q_{sub}	Ultimate soil shear resistance of cylinder of diameter B_b and length down to underream, kips (pounds)
Q_{sud}	Downdrag, kips (pounds)
Q_{sui}	Ultimate skin friction resistance of pile element i, ksf
Q_{sur}	Ultimate skin resistance, kips (pounds)
Q_u	Ultimate bearing capacity force, kips
Q_{ug}	Ultimate load capacity of pile group, kips
$Q_{ug,lower}$	Bearing capacity of base at top of lower (weak) soil, kips
$Q_{ug,upper}$	Bearing capacity in the upper soil if the softer lower soil were not present, kips
Q_{up}	Uplift force on foundation, kips
Q_w	Working load, kips (pounds)
Q_{DL}	Dead load of structure, kips (pounds)
R	Resultant load on foundation, $(Q^2 + T^2)^{0.5}$
R_{bc}	Scale reduction factor for end bearing capacity in clay
R_{bs}	Scale reduction factor for end bearing capacity in sand
R_d	Ratio of equivalent embedment depth to footing width, D_e/B
R_e	Eccentricity adjustment factor
R_k	Bearing ratio using CPT procedure for estimating bearing capacity
R_v	Strength reduction factor of vane shear test
S	Average penetration in inches per blow for the last 5 to 10 blows for drop hammers and 10 to 20 blows for other hammers
S_r	Spacing between vertical reinforcement rods in soil, ft
S_s	Shape factor, assume 1.000
S_u	Depth of scour, ft
SPT	Standard penetration test
T	Horizontal (lateral) load on foundation, kips
T_a	Allowable lateral load capacity, kips
T_u	Lateral load capacity, $T_{us} + T_{up}$, kips
T_{ug}	Lateral load capacity of pile group, kips
T_{ul}	Ultimate lateral load capacity of long pile in cohesionless soil, kips
T_{up}	Lateral load pile capacity, kips

Symbol	Description
T_{us}	Lateral load soil capacity, kips
T_v	Torque of the vane test, kips-ft
W	Lateral length of a foundation, ft
W_p	Pile weight or pile weight including pile cap, driving shoe, capblock and anvil for double-acting steam hammers, kips
W_r	Weight of striking parts of ram, kips
W'_T	Total effective weight of soil and foundation resisting uplift, kips
W'	Effective lateral length of a foundation, $W - 2e_W$
Z	Section modulus I_p/c, ft^3
Z_a	Depth of the active zone for heave, ft
Z_c	Concrete section modulus, ft^3
Z_{st}	Steel section modulus, ft^3
α_a	Adhesion factor
α_f	Dimensionless pile depth-width relationship factor
β	Angle of ground slope, deg
β_f	Lateral earth pressure and friction angle factor
γ	Wet unit soil weight, lbs/ft^3
γ_c	Moist unit weight of weak clay, $kips/ft^3$
γ_{conc}	Density concrete grout, $kips/ft^3$
γ_d	Dry density, $kips/ft^3$
γ_p	Pile density, $kips/ft^3$
γ_s	Unit wet weight of sand, $kips/^3$
γ_{sand}	Unit wet weight of the upper dense sand, $kips/ft^3$
γ_w	Unit weight of water, 0.0625 $kips/ft^3$
γ_D	Unit wet weight of surcharge soil within depth D, $kips/ft^3$
γ_H	Wet unit weight of subsurface soil, $kips/ft^3$
γ_{HSUB}	Submerged unit weight of subsurface soil, $\gamma_H - \gamma_w$, $kips/ft^3$
γ'	Effective wet unit weight of soil, $\gamma - u_w$, $kips/ft^3$
γ'_b	Effective wet unit weight of soil beneath base, $kips/ft^3$
γ'_c	Effective wet unit weight of clay, $kips/ft^3$
γ'_s	Effective wet unit weight of sand, $kips/ft^3$
γ'_D	Effective unit weight of soil from ground surface to foundation depth, $kips/ft^3$
γ'_H	Effective unit weight beneath base of foundation to depth $D + H$ below ground surface, $kips/ft^3$
γ'_L	Effective wet unit weight of soil along shaft length L, $kips/ft^3$
Δ	Differential movement within span length L, ft
ΔL	Pile increment, ft
δ	Angle of base tilt, deg
δ_a	Soil-shaft effective friction angle, deg
ζ_c	Dimensionless correction factor related with cohesion accounting for foundation geometry and soil type
ζ_{cs}	Dimensionless correction factor related with cohesion and shape
ζ_{ci}	Dimensionless correction factor related with cohesion and inclined loading
ζ_{cd}	Dimensionless correction factor related with cohesion and foundation depth
$\zeta_{c\beta}$	Dimensionless correction factor related with cohesion and ground slope
$\zeta_{c\delta}$	Dimensionless correction factor related with cohesion and base tilt
ζ_{cp}	As above except for piles
ζ_γ	Dimensionless correction factor related with soil weight in the failure wedge (repeat as above for factors s, i, d, β and δ)
$\zeta_{\gamma p}$	As above except for piles
ζ_q	Dimensionless correction factor related with surcharge (repeat as above for factors s, i, d, β and δ)
ζ_{qp}	As above except for piles

Symbol	Description
θ	Angle of resultant load with vertical axis, $\cos^{-1}\left[\frac{Q}{R}\right]$
λ	Lambda correlation factor for skin resistance of Vijayvergiya & Focht method
ρ	Settlement, inch
ρ_b	Base displacement, inch
ρ_{bu}	Ultimate base displacement, inch
ρ_e	Elastic pile settlement, inch
ρ_i	Immediate plate settlement, inch
ρ_u	Ultimate pile settlement, inch
ρ_z	Vertical displacement at depth z, ft
σ'_d	Effective soil or surcharge pressure at foundation depth D, $\gamma'_D \cdot D$, ksf
σ_{ho}	Total horizontal in situ stress, ksf
σ'_i	Effective vertical stress in soil in at shaft (pile) element i, ksf
σ'_m	Mean effective vertical stress between the ground surface and pile tip, ksf
σ_n	Normal stress on slip path, ksf
σ_o	Reference overburden pressure, 2 ksf
σ'_p	Maximum past pressure in soil, ksf
σ'_v	Effective vertical stress, ksf
σ_{vc}	Total vertical pressure in soil including pressure from structure loads, ksf
σ'_{vc}	Effective total vertical pressure in soil including pressure from structure loads, ksf
σ_{vo}	Vertical overburden pressure, ksf
σ'_{vo}	Effective vertical overburden pressure, ksf
σ'_z	Effective overburden pressure at the center of depth z, $0 < z \leq L$, ksf
σ'_L	Effective soil vertical overburden pressure at pile base, $\gamma' \cdot L$, ksf
$\sigma'_{L/2}$	Effective stress at half the pile length, ksf
τ	Shear stress, ksf (psf)
τ_{max}	Shear stress at failure, ksf (psf)
τ_s	Soil shear strength, ksf
τ_u	Field vane undrained shear strength, ksf
υ_s	Poisson's ratio for soil
ϕ	Angle of internal friction of soil, deg
ϕ_{sand}	Angle of internal friction of upper dense sand, deg
ϕ'	Effective angle of internal friction of soil, deg
ϕ_a	Friction angle between foundation base and soil, deg
ϕ_g	Friction angle of granular material, deg
ψ	Angle of shear zone failure with respect to foundation base, Figure 1-3, $45 + \phi'/2$, deg
ω	Angle of pile taper from vertical, deg

INDEX